FROM THE GROUND UP

FROM THE GROUND UP

REBUILDING GROUND ZERO TO RE-ENGINEERING AMERICA

AMBASSADOR CHARLES A. GARGANO

WITH IAN BLAKE NEWHEM

PHOTOGRAPHY BY GARY MARLON SUSON

Post Hill
PRESS

A POST HILL PRESS BOOK

From the Ground Up:
Rebuilding Ground Zero to Re-engineering America
© 2019 by Charles A. Gargano and Ian Blake Newhem
All Rights Reserved

ISBN: 978-1-64293-143-3
ISBN (eBook): 978-1-64293-144-0

Cover art by Rahul Panchal, storyandpromise.com
Photos by Gary Marlon Suson and Lester Millman
Interior design and composition by Greg Johnson, Textbook Perfect

Post Hill Press
New York • Nashville
posthillpress.com

Published in the United States of America

"Far better it is to dare mighty things,
to win glorious triumphs,
even though checkered by failure,
than to take rank with those poor spirits
who neither enjoy much nor suffer much,
because they live in the gray twilight
that knows not victory or defeat."[1]

—THEODORE ROOSEVELT

Contents

Foreword

For all his polarizing rhetoric in the run-up to Election Day 2016 and beyond, President Donald Trump hit at least one nail squarely on the head: Our country really is a disaster in need of recovery. Regardless of our disparate political leanings, the majority of Americans simply know in our bones that we're in real trouble. The mandate Americans gave Trump highlights the crises that are threatening to break us economically, politically, spiritually, and *literally* if we don't start immediately improving our infrastructure, solidifying our social fabric, and surgically restoring the ideological and cultural vertebrae of our nation's spine, which has kept us standing for two and a half centuries despite a lot of hard hits.

Sure, we've undergone tough times before, but today an unprecedented barrage of calamities has coalesced into the perfect storm, a megadisaster that will either swamp us once and for all or offer an opportunity for us to re-engineer and rebuild from the ground up. I believe we will succeed.

We have to.

But while some things are working better than ever in America today, business as usual—especially over the past couple of decades—is killing us. A lot of people expect opportunities to fall into their laps. When that doesn't happen, the only thing they're not too lazy to do about it is complain. Sure, they'll take to social media, they'll block traffic and break windows, they'll demonstrate on our highways and in our subway stations. Those with means might donate to their favorite agitators. Of course, they'll act surprised when nothing really changes.

At the same time, we've got politicians who are models of inaction, blathering a bunch of hot air to help their re-election bids and nothing more, selling out the very people whose votes they depend on. We even suffer some of the business community's adding fuel to the fire when it refuses to accept having to pay its fair share for the resources it uses. We have executives who will destroy a company to make short-term profits and leave a complete mess for the next guy to clean up. Not only is it dysfunctional and dishonest, it's just bad business.

So the question we have to ask is: what do we tackle first in the inner cities? The national debt? Partisan politics? Race relations? Income inequality? How about that disintegrating infrastructure?

Make no mistake. This moment marks the biggest existential threat we've faced since 9/11. Maybe even since the Civil War and the Great Depression. Our country needs complete re-engineering from foundation to roof. We need a plan—a prescription for massive national evolution, one step at a time.

My friend and colleague Charles A. Gargano has a plan to do just that. In *From the Ground Up*, one of the country's most accomplished civil engineers—the man principally responsible for the revival of New York following the terrorist catastrophe of September 11, 2001—outlines transformative instructions for massive national evolution, one step at a time. Using as a model the multifarious challenge of turning Ground Zero from a "pit and pile" into an international emblem of resurrection, Gargano proposes a series of incremental actions for us to take in order to salvage, rehabilitate, and rebuild a broken country, all within the framework of "re-engineering."

As a laborer, civic leader, fundraiser, politician, and professional engineer, Gargano is perfectly positioned to recommend in honest, plainspoken, and often controversial prose his revolutionary but relatively simple and sensible solutions to our crises. He shows us how to turn the disaster scene our country risks becoming into a beacon of brilliance for which we can once again feel pride and how to renew our role as true world trailblazers. Just as the "Freedom Tower" slowly rose from the ashes of Ground Zero despite near-impossible odds, our country too can rise again. This is the best plan I've seen yet to make that happen.

As chairman and CEO of the Empire State Development Corporation, commissioner of the Department of Economic Development, and vice chairman of the Port Authority of New York and New Jersey during America's darkest days and thorniest re-engineering project, Gargano was a classic inside man from day one. On the site, he and I watched the second plane hit. We met up with Mayor Rudolph Giuliani within minutes. Gargano was there during every stage of the rebuilding of lower Manhattan, orchestrating innovative developments as he had done for years with the successful revamping of 42nd Street, Harlem, and Niagara Falls, as well as many other projects that changed the face of the city and state of New York. He was there in all the meetings with me, Giuliani, developer Larry Silverstein, architects David Childs and Daniel Libeskind, and other top city, state, police, FBI, FEMA, and Port Authority officials. In short, Charles Gargano is one of only a handful of people who personally understand the intricate challenge of that undertaking and who had a hand in all of it.

Because of this "in the trenches" experience, on top of his expertise as a political operative and major Republican donor/fundraiser, ambassador, community leader, and civil engineer, Gargano has become a pioneer and thought leader in the specialized field of re-engineering. Re-engineering is the only way to capitalize on our strengths and fix brick by brick, from the ground up, a system at risk of becoming ruins. Gargano has a reputation for turning rubble into gold.

The presence of One World Trade Center, the tallest structure in the Western Hemisphere, is no accident. It represents no less than an epochal and triumphal event, defiant and life-affirming, a signature moment of grace in the tortured early history of the twenty-first century. It was never going to be easy. Using each major lesson learned during the decade-plus of rebuilding the 9/11 site—each achievement and setback—*From the Ground Up* offers Americans for the first time a comprehensive, practical, and inspiring plan for re-engineering the entire country.

It's about time.

—GOVERNOR GEORGE E. PATAKI
NOVEMBER 2018

Introduction:
Recall and Analyze

The Greatest Generation

I hate to sound like an older man, but I often long for the "good old days." In thirty seconds, I can tell you why. Not because I miss "Pee Wee" Reese and Jackie Robinson, John Wayne, Ed Sullivan, and Frank Sinatra, though I do. No, it's because in the '40s, the Greatest Generation punched the Nazis right in the nose. In the '50s, Eisenhower stretched a gleaming interstate highway system forty-two thousand miles from sea to shining sea. In the '60s, young JFK promised we would launch humans to the moon, and, astoundingly, before the '70s began, humanity took that giant leap into a new and thrilling era.

At the turn of the nineteenth century, my own grandfather, an immigrant from Sant'Angelo dei Lombardi in the Avellino province of southern Italy, stepped anxiously along with millions of other émigrés from Italy, Ireland, Asia, and Eastern Europe through the gateway of a red brick castle on Ellis Island. Destitute, hopeful, hardworking, and ambitious, these people would design and build all the bridges, tunnels, subways, and skyscrapers that still define my city—America's city—New York.

Your city, too.

As a people, we're no strangers to such Herculean engineering projects. Starting centuries ago, we laid out the rails—more than a hundred billion pounds of steel—for train travel. From east to west, city after city rose from the desolate and unforgiving landscape—Boston, Pittsburgh, Indianapolis, Chicago, Kansas City, Denver, San Francisco, and everywhere in between. Towering dams. Half a million miles of electrical transmission lines. Immense aqueducts, "water tunnels," like the one in New York with a storage capacity of 550 billion gallons. That one system delivers 1.2 billion gallons of fresh water daily, 95 percent of it progressing by gravity alone[1]—technology perfected in the Assyrian empire in the ninth century BCE.[2]

It took true grit, vision, and ingenuity to undertake such monumental tasks. The same "right stuff" required by Frederick Law Olmsted when he looked at swampland in the middle of Manhattan and built, from the ground up, an eight-hundred-acre idyllic retreat—America's most visited urban park—for the public's health and recreation. Without the respite I have felt there, there were days—especially in the fall of 2001—I might have gone crazy. The same "right stuff" that appeared again in 1968, when the builders of the first World Trade Center (WTC) site had to figure out what to do with the 1.2 million cubic yards of rock and dirt they'd excavated for the foundation. Should they just heap all that crap into the harbor, or jam up local landfills?

Instead, in concert with David Rockefeller's urban renewal mission, they conceived the idea of using all that fill material to expand the Manhattan shoreline across West Street to enlarge the city itself. That's how Battery Park City, a seven-hundred-foot, six-block, ninety-two-acre add-on, arose as though straight out of the Hudson. As an ancillary benefit of building the world's tallest towers at the time, we created all that housing space. We also built the bucolic 1.2-mile riverfront esplanade and more than thirty acres of new parkland for guys like me to admire on tumultuous days.

And speaking of which, more recently, we found ourselves having to reconstruct much of the downtown of America's biggest and most recognizable city after terrorists annihilated its heart in one unimaginable attack. That's where I come in, but more on that later.

Suffice it to say it took the vision of the nation to make that happen. Consider again that audacious "moon shot." Or Edison's and Tesla's competing wonders of modernity. Imagine what it must have been like for the citizens of Cleveland, Ohio or Wabash, Indiana, to see their cities "lit like midday" by electric arc lights in the 1880s. Or how about watching the Roebling couple's masterpiece erected between Manhattan and Brooklyn, over which 120,000 cars pass now every day? On top of which, if you look carefully as I do, you'll see peregrine falcons nesting.

Somebody had to dream these things, necessity often being the mother of invention. Only then could the likes of Washington Roebling and his wife, Emily—the first female field engineer—start building caissons and stringing cable. Only then could we band together, roll up our sleeves, and get to work under skillful leadership. We had to gather the facts, truly understand the problem. We had to take the long view like those falcons get when they wheel above the city. We had to triage our projects, not get mired in the trivial. Along the way, we had to overcome obstacles and let criticism roll off our backs—or understand that our detractors were right after all, and we had to return to the drawing board having learned from our mistakes. We had to adapt to unforeseen glitches in our plans. We had to compromise.

We Americans might be distinctly talented and proficient in these areas. We were "Jersey Strong" after Superstorm Sandy. "Boston Strong" after the marathon bombings. New Orleans's recovery from Hurricane Katrina's devastation, albeit slow and imperfect, took the combined strength of a truly united nation to pick up the pieces and start again— and we learned the hard way that we can't always rely on our leaders to direct us. We're so good at rebuilding after utter destruction that we've even done it for our former enemies, in Japan and Germany, and, of late, in Iraq and Afghanistan.

How did we accomplish all these miracles of leading-edge engineering, despite the overwhelming odds? Mere elbow grease and the right tools in our belts? Trial and error? Confidence, in both senses of the word? Faith? Fearlessness in the face of risk? A service mentality? Sure, all those things were vital. But they weren't the main thing. No, we could never have soared to the heights and tunneled straight through every obstruction had we lacked a common *vision*, a shared mission. Vision and leadership plus determination and know-how—the right team following a sensible and well-articulated plan—allow us all to realize the American Dream.

How else can you explain how a thirteen-by-two-mile island of bedrock and wildlife sold for a song to colonists by the Lenape Native American tribe evolved into the country's biggest city by far? Brooklyn alone—my hometown—if split from the other boroughs would still be the fourth largest city in America. Now put New York into the pot with all the other big cities and little towns in the four thousand miles between Miami, Florida, and North Pole, Alaska. In a mere nine generations, we grew from three million to more than three hundred million. Along the way we evolved from scrappy tenants to commanders of the freest, richest, and most technologically advanced nation that has ever existed.

This didn't just "happen." It happened because we believed it could happen, and we backed up our belief with discipline, chutzpah, and the American work ethic. And we had leaders who laid out plans and sold them persuasively to the public. Guys like Eisenhower, Reagan, and Pataki. Or let's go back further to the founding fathers. Talk about vision, leadership, and audacity. They really saw us this far ahead, at our best.

And Now for the Bad News

Despite our former glory, we're in real trouble today. Sure, some important aspects of our lives have improved dramatically over the past half century: life spans, working conditions, and civil rights, to name just a few. We still build huge, iconic buildings like 30 Park Place and 10

Hudson Yards. We innovate and dominate in tech. We're still the envy of the world in many ways.

But what if we honestly inventory the magnificent feats we've achieved in the distant and recent past and contrast them with our current state? We find our country is, in fact, a relative disaster, just like the man says. Anyone who lands at LaGuardia or one of America's other "third-world" airports or drives to the Bronx from JFK airport knows this intuitively (thank God and our government that both airports are under renovation). Anyone who watches ten minutes of cable news knows we're buried up to our necks in acrimony and meanness, and many of our leaders have lost their way—unless it's the way to the bank vault or federal prison. Anyone who tries to get on a plane in a hurry, obtain a contractor's license or a building permit, or get untangled from other sticky red tape knows we've got too much government—and it's focused on all the wrong things.

Anyone is you. Anyone is me. We both know that, on the whole, our health care and education systems are straggling like a two-legged puppy on a greyhound track. Our health care efficiency ranks in the bottom tenth of nations,[3] and kids in half the countries of the world lead ours in math, reading, and science.[4] Since when are we okay hanging out in the middle of the pack? Especially when we both know we're under attack for our "exceptionalism." Chances are diminished for us locally, yes, but we haven't seen the last of terror arising on alien shores, nor of subversive agitation born and raised in our own heartland. Maybe worst of all, we can't go to the mall, take our kids to the park, or even turn on the TV without witnessing daily the exponential erosion of our traditional values. It doesn't help the situation—our standing, our direction—when at the top we find a man like Donald J. Trump.

So our physical infrastructure and social, cultural, and political stability are all badly in need of buttressing—or a gut renovation. The vast majority of us agree the country's headed in the wrong direction.[5] We're right, and admitting the problem is the first step in overcoming it. But it's just the first step. We have to re-engineer the whole country from the ground up—its social fabric, its economy, its politics, and its physical plant. That's what this book is about.

The good news is there is good news, points of light shining through the darkness. You are the lightkeeper. The architect of our future. You're the engineer, the builder, and the end user. You're the foreperson. The steward. The cure.

Look, I know that all sounds grandiose. You might be feeling bleak and frustrated and ready to turn in your key, curl up in a corner. But ask yourself what attracted you to this book. Maybe you're not quite ready to count out the extraordinary progress we've made and will continue to make if we stay positive and reorient our ways where it's obvious we've gone astray. The truth is, you will not be able to count on a single other human to even start this project, much less finish it. The only person you absolutely know can get a grip on this off-course country and help steer it true is you.

I want to help, and I've got some hard-won experience that ought to come in handy. I ask you to lend me your trust for a little while. I promise I won't waste your time.

The Pit and the Phoenix

So how do we really rise? Where the hell do we start this national re-engineering project? I get it. It's intimidating. Collectively, we've got a list of, say, a dozen or so major calamities occurring simultaneously, including a broken body politic, a cultural calamity, lagging leadership, a disintegrating infrastructure, and a lack of funds to execute any truly big ideas. And we can't just blame the government—we, the people, are responsible. We have some work to do, too. We're rude to each other, often uncaring. We steal from our own brothers and sisters. We allow sloppiness to pollute our environment. Some of us demonstrate unpatriotic ideas and actions daily.

So how are we supposed to triage those problems, prioritize them in order of importance and worthiness of our time, money, and human resources? And how exactly do we get to work once we have prioritized?

Don't we have to bulldoze a lot of crap away before we uncover solid ground upon which to rebuild? Finally, what meaningful part can any individual play in soothing our nasty national wounds?

Surely deciding all that based on our own party's ideology has not worked so far. Basically, we're frozen by the knowledge that things are getting worse. Look around. Nobody's minding the store. No, I'm not a pessimist by nature. I'm a realist. There's no room for Pollyannas on our planning board, nor do we want Debbie Downers at the table. We have to first believe we can pull ourselves out of these doldrums. Then we have to act—and act fast—to untangle the knottiest of these problems, to begin to debride and heal the worst of our festering lesions.

So what if there were a model out there, an example of how we came together to resolve some massive challenge in relatively short order? It would have to be really big. And preferably contemporary. It would have to provide a paradigm for all the steps needed to solve major difficulties. Basically, we'd need the exemplar instance of converting rubble into gold, of turning a pit and pile into an emblem of accomplishment against impossible odds. We'd need evidence that we *can* move mountains—because we literally *have*.

Such an example towers over our most populous and visited city. After September 11, 2001, we managed, albeit often contentiously and never without struggle, to rebuild on holy ground an icon not only of survival but of triumph. Of freedom. As the governor says in his foreword, I was there every step of the way. I don't recount that fact out of any sense of pride. I wish to God I never had to do all that. I wish those assholes who attacked us had never been born.

No, I tell you this because, since that time, I've thought long and hard about the intense process we went through to turn a mark of victimhood into a symbol of prosperity and defiance. If you'll allow an old man a modicum of pride, it could have been my greatest legacy to have served on the team that saw that new tower finally, boldly erected. But ever since the first blush of excitement died down after I saw the 1,776-foot emblem of our success opened to the public on November 3, 2014, something has nagged at me.

I've thought about that rebuilding. I've thought about all our hopes tempered by grave disappointments, all the serpentine roads we had to travel, and all the collateral damage—especially in human terms.

And I've finally realized that what has been bugging me is this: we're not done yet re-engineering.

I'm in my early eighties. Most of my contemporaries have long since retired. They're on golf courses and beaches, enjoying the fruits of their considerable labors. I wish I could rest, spend more time with my grandkids, maybe travel more.

But I can't.

I can't because I keep thinking about all the metaphorical—and sometimes literal—Ground Zeros still in need of cleanup. I believe the exploit that began on September 11, 2001 and ended more than a decade later provides the very template for solving the difficult problems threatening our survival as a nation.

If we examine the steps we took and the lessons we learned turning Ground Zero into a beacon of hope and possibility, we can apply them to some of the grimmest challenges America has ever faced. Where we made mistakes in that project, we can learn and do better.

You'll notice that the governor and I have used the term "re-engineering" several times already. Let me explain what we mean. To overcome our problems, we'll need to hone more than our rethinking, retooling, and rebuilding skills. A licensed professional engineer—the role I studied and practiced for the past sixty years as a career—devises, develops, and sustains all the movable parts in a major project, both individually and holistically. A momentous engineering project requires fact-gathering, solid leadership, hierarchy, prioritizing, collaboration, planning, nerve, adaptability, compromise, a thick skin, straight talk, execution, and, of course, fundraising.

I contend there isn't an obstacle we face politically, economically, socially, or structurally that we can't lick by adopting this re-engineering model, as long as we thoroughly grasp each of those steps in some detail. And, naturally, as long as we really understand the nature of the problems we need to tackle.

Take Stock

Nothing prepared me for "the Pile" the governor and I inspected the late morning of September 11, 2001.

A little after 9 a.m. that gorgeous Tuesday, Pataki and I had watched United Flight 175, a Boeing 757, slam into floors seventy-seven to eighty-five of the southern façade of 2 WTC, the South Tower, at 590 miles per hour. The Fire Department of New York (FDNY) had been coordinating the evacuation of both towers ordered by the police from my agency, the Port Authority.

We were headed to the city's emergency command center on the twenty-third floor of 7 WTC. It had been only seventeen minutes since terrorists used a Boeing 767, American Flight 11, as a 470-mile-per-hour missile to obliterate floors ninety-three to ninety-nine of the North Tower, 1 WTC.

That one might have been an awful accident. But this second one—now it was war.

A quick-thinking national ops manager for the Federal Aviation Administration (FAA)—his first day on the job—had already ordered a regional ground stop, and a few minutes later, he took the unprecedented step of grounding all civilian aircraft in the United States. By 9:45 a.m., the FAA had shut down all U.S. airspace. Governor Pataki and I were whisked up to Gramercy Park to meet Mayor Giuliani at the thirteenth precinct on East 21st Street. Rudy and George both spoke with the president, who by that time was flying back to Washington on Air Force One from a speaking engagement at a school in Sarasota, Florida. He cautioned us that more attacks were expected.

By that time, both towers had tumbled, the Pentagon had been savaged, and United flight 93 had gone down in a ball of flames into a field outside Shanksville, Pennsylvania. Tower 7 had suffered grievous injury by debris from Tower 1's collapse, and ten floors on its lacerated west side were on fire. The water lines had been demolished by the Twins' downfall. Left to nature and chemistry, the fires would either burn themselves out or consume the whole structure.

In either case, structural steel is not supposed to literally melt. Before September 11, no building higher than fifteen stories had ever collapsed because of fire.[6] Already that day, there had been two exceptions. Steel melts.

There were nonstop reports of hijacked planes still whizzing toward American targets. The mayor ordered the evacuation of Lower Manhattan at around 11:00 a.m., and he headed to the site on foot. I hunkered down the rest of the day with the governor, coordinating and receiving stunning, unimaginable reports on loss of life, site damage, and the state of the nation. People reported seeing the towers burning from fifty miles upstate. The look on the faces of the officers and police officers as they entered our command center from the disaster scene jarred me in a way I'd never experienced. It was all I could do to focus on my job, but focus we all did. By 12:10 p.m., U.S. airspace had been cleared of all four thousand civilian aircraft that had taken off that morning, and everyone felt a little safer.

A little.

Later that day, Pataki and I re-entered the site. That's when what was left of Tower 7—home of the Secret Service, the CIA, and our own Office of Emergency Management (OEM)—fell, fully engulfed in fire. It was 5:20 p.m. Even as I watched that forty-seven-story building collapse, I was remotely aware that the moment would be relegated to a mere footnote in the story of that day. It's amazing how quickly we humans can adapt on the fly and still rise to the occasion. Or maybe it was just shock.

In front of that burning nightmare were the remnants of what had once been the world's tallest towers. Their combined 220 stories lay compacted and smoldering over sixteen acres, encased in a profuse storm of acrid dust—charred, pulverized, and carbonized concrete, glass, gypsum, and a toxic fusion of carcinogens and other poisons. There were hundreds of thousands of tons of debris, including 185,101 tons of steel. There were two vaporized passenger jets—only one complete jet engine would survive. And worst, of course, were what we assumed could be tens of thousands of victims, including hundreds of first responders. More than fifty thousand people worked in the Twin Towers.

I agree with President George W. Bush that it looked like a war field. Some companies, like Cantor Fitzgerald, lost two-thirds of their workforces, and 2,753 families suffered an immeasurable, incalculable, inexcusable loss in less than two hours.

The site was apocalyptic—terrifying, depressing, humbling, infuriating, and exceptionally intimidating to look at and walk upon. But within minutes of the collapse of the Twin Towers, firefighters, cops, military and government personnel, and citizen volunteers began risking their lives to comb through that wreckage for signs of life. They would work twenty-four hours a day for a week and pull only twenty survivors from the pile, including three of my own—two Port Authority police officers and one of my structural engineers. For 230 days thereafter, workers kept searching for body parts, finding and cataloguing nearly twenty thousand. They uncovered 291 intact bodies.[7]

For the next decade and then some, it was hard to look at that space. From a distance, especially, such as looking down Sixth or Park Avenue, the absence of the towers strangled my heart, as it did for all New Yorkers and Americans. You could feel it palpably, like you had just had your two front teeth viciously kicked out. But close up, for the twenty months during which workers reduced that pile in five-pound buckets of debris at a time, it remained a dense, gnarled, seemingly impenetrable mess.

The fires would rage for ninety-nine days. The reek of human flesh and toxins stung the back of the throat and choked the lungs in a way that perhaps only combat soldiers and the extremely unlucky had previously known. Many would suffer devastating health consequences years later, raising the death toll significantly and tragically.

Taking stock of that site was soul-shattering. As vice chairman of the Port Authority as well as Pataki's chairman and CEO of Economic Development, I considered it *my* mess. My first impression every time I stared down that site was, "All is lost." I couldn't help but feel that way—overwhelmed, stricken. Then that gloom of hopelessness always made way for pinpricks of light, like flashlights through the fog of a battlefield. One light was the support of our president—the light of patience, prudence, and resolve. Another was the light of the troops—official and unofficial—volunteering to risk their lives for others and for our country. When all those beams of light came together over the hours, days, weeks, months, and years, the mission was as clear as the weather that Tuesday. No one had to type it up in a memo or press releases: We had to clean up. We had to help families. We had to rebuild.

But the sight of it all. The smell. The immensity. It twists the human soul still, the way those extreme forces warp steel beams. *All is lost.*

Survey Says...

I began my career in the mid to late '50s as a site engineer, part of a four-man surveying team working for the city of New York, Brooklyn borough's engineering department. We were developing Canarsie and Mill Basin, which were dumps at the time. My job was to operate the sensitive instruments. I had to take exact measurements of every inch of the sites to determine boundaries and assess the possibilities and limitations of the locations where the city planned to build—mostly schools. I took stock of the shape, contour, and other features of a site before any engineering or construction could take place.

For instance, the plan for building a junior high school might call for clearing a certain area of trees and rocks, with the goal of saving X number of them for aesthetic, legislative, or counter-erosion purposes. I had to stake out sites with orange markers so the bulldozer operators would know what to clear and what to keep. I also had to be sure what was beneath that ground. The last thing you want when digging out tons of earth is to be surprised.

So there I was, in my twenties, staking out the proposed layouts of buildings in New York City and on Long Island, as close to absolutely accurate as possible. That's how the concrete workers would know where to pour the foundation so all the aboveground stuff would line up and not topple under its own weight. Then I staked the roads for the paving and curb work crews.

Planning is everything. I learned that in engineering school at CUNY and Fairleigh Dickinson University where I received an MBA, and later during my master's program in civil engineering at Manhattan College. But you can't do any planning unless you know what you're working with. You have to know your materials, tools, skills, and the conditions on the ground at your building site.

I learned in the first two years about the strength of materials, physics, a lot of math, and so forth. Then I got into structures such as retaining walls, columns, and footings. I learned how to design and build them from the ground up, to hold a massive amount of weight on their shoulders. Eventually, I learned how to build bigger and more complex structures, and then to make them beautiful. But even at the start, the teachers stressed the importance of planning properly and calculating precisely.

My son Larry studied architecture, and the foundations of that discipline are similar. They include the designing of structural steel; civil engineering of sewers and power grids; the designing of load-bearing, connecting beams based on the forces the spans can endure; and even disaster remediation—the basis for such computations are the same for architects, engineers, and builders. It doesn't matter whether you want to build the tallest supertower in the country or rebuild the whole country from its footings on up.

But none of it can ever see the light of day without a careful and correct site survey. The way you build a skyscraper in the sands of Dubai differs dramatically from the way you build on the banks of the Hudson River. And methods of cleaning up the consequences of 60,029,283 cubic feet of steel, glass, Sheetrock, office furniture, and human remains collapsing is different from the way you eradicate the results of, say, a tsunami.

Why do I mention all this? Well, the same processes—surveying, designing, calculating, connecting, structuring—are all necessary for re-engineering and building your ideal life and our ideal country. What's your location? Where are your boundaries? What are the features and restraints you have to work with? What are the weak links and bottle-necks of any previous incarnation or the ground itself on which you plan to build? Does the site need cleanup or other preparation before you start to redevelop? What's the purpose of this project, whom will it serve, what will it look like, and how will it interact with its environment?

Why those functions? Just a whim? Or is there some significant utility, a real need? How can you capitalize on the history, geography, culture, and natural resources of your site? All of that goes into designing, engineering, and building physical structures as well as human lives and nations.

So let's start our survey of America's Re-Engineering Project. We'll hark back to some lessons from Ground Zero and earlier in my surveying, engineering, political, and economic experiences. I guarantee you an exciting—if bumpy—ride.

1

Identify and Prioritize

Get Real

Impossible? Too hard? Too painful? Too controversial? Maybe just throw some tarps over the whole American jumble and sneak away whistling "Dixie," pretending it'll all work out? That's tempting. And there's good precedent for it. But history doesn't favor those granted the opportunity to survey a situation but who look the other way. About thirty miles from my ancestral home in Italy, two thousand Pompeiians lost their lives in 79 CE because they wined and dined and went about their days ignoring the obvious "rumblings" of Vesuvius. The Gulf of Naples literally boiled before them, their wells dried up, and millions of Roman rats decamped for parts south—and the citizens ate their dates and honey. History lies littered with the fossils of civilizations that perished in an instant of inattention—or after years or even centuries of obvious decay that no one took seriously enough, that no one felt personally responsible for.

Two thousand years later, it took the American Allies far too long to get into World War II—and, in fact, several red flags waved in front of our complacent faces before Pearl Harbor. But imagine if we hadn't gone at all and just hoped the world would sort itself out on its own. Why on earth would we commit our precious blood? All the evidence showed that Hitler and his blitzkrieg buddies aimed to dominate the world from their exponentially expanding Axis lands. Japan invaded China. The Nazis overran Poland. Then a major morning strike caught us unawares on our own soil.

But of course we had to respond. Millions of men and women considered it not only their national duty but their personal responsibility. They knew what they were facing, and they went anyway. The point is, a survey requires a few key things, not the least of which are open eyes, courage, and honesty. But it all comes down to people's own sense of accountability. Even against the odds, you can empower yourself to believe that your attitudes and actions can make a difference in the big picture. They can.

Let's get to work.

The Big Ten

Let's conduct a survey of our country's biggest problems. Let's look at our national "Pile." We have to know the current conditions before we can contemplate re-engineering. Maybe your list includes some items different than mine. That's what makes our country the diverse place it is. But I expect there will be lots of overlap. These are in no particular order.

Critical Infrastructure

It's aging—in many cases, obsolete—unmaintained, and inefficient. Commuters on the FDR/Harlem River Drive feel this fact jarring their spine. Did you know that Americans take more than two hundred million trips *daily* across deficient bridges in the nation's 102 largest metropolitan regions?[1] If you're unfortunate enough to live in Flint, Michigan (or one of the hundreds of other localities with dangerous drinking water), you can taste our disintegrating infrastructure. If your life and livelihood depend on the efficacy of one of our one hundred thousand miles of levees, you'll remember the Ninth Ward catastrophe of 2005 and forever sleep with one eye open.

The evidence of the decline we let happen surrounds us. Recently I drove to a meeting at George Pataki's office on Sixth Avenue from my summer house in Bridgehampton on Long Island. Needless to say, this is no longer the pleasant drive my parents used to take with us kids— my brother Frank, my sister Connie, and me—in the '40s. Back then, the South Shore neighborhood where we spent our vacations still had mom-and-pop stores, dirt roads, and cows.

First, without traffic—or in a driverless car!—this modern-day trip to Pataki's office should have taken two hours. But who remembers what this strange thing called "without traffic" means? The drive took twice that long on the choked Long Island Expressway. There's an old joke that if a groundhog farts in the granny lane of the LIE, a three-hour shutdown will surely ensue, attended by all manner of state and county agencies, on our dime. That day, I passed two emergency pothole repair crews and two people with obvious tire problems from potholes the size of the Grand Canyon.

Is it any wonder that commuting to work ranks as the least pleasant daily activity the average person experiences? It's one of the major reasons road rage is on the rise. Eighty percent of us confess to expressing it within the past year—and that number is growing.[2] The other 20 percent are full of shit. Americans sit in traffic for a combined eight billion hours a year. That's nearly a *million years*. A million years of productivity, family time, and leisure time choked by diesel. In our ten most congested cities—you know who you are!—individual commuters spend up to two full workweeks in gridlock every year.[3] The cities experiencing the greatest growth experience the worst traffic. The more traffic, the greater the abuse of the infrastructure.[4]

In 2015, Congress passed a $305-billion five-year highway and transit bill. That might sound like a lot of money, but it falls far short of the amount needed to bring our highways, bridges, and tunnels up to speed. In one recent year, the U.S. spent $228 billion on transportation and transportation safety.[5] But we suffer an $836 billion backlog of repairs and improvements to our roads and bridges. We need a further $90 billion to clear the logjam of public transportation systems, according to the Federal Highway Administration (FHA).[6] That doesn't include building new roads to ease traffic, or updating the sewers and canals running beneath the roads to twenty-first-century standards.

Total government spending on infrastructure is at a thirty-year low just when we need it most.[7] Now our president plans to raise military spending by 10 percent. Where do you think that money's going to come from? Trump's plan to parley $200 billion in federal money into $1.5 trillion for reviving America's infrastructure (by leveraging tax dollars and private investment)[8] is a fantastic start, but just a start. Is it really feasible to raise those funds? And how exactly will they be distributed? Although both major political parties have failed us here, the Republicans have historically been less likely to fund infrastructure projects—for reasons that entirely escape me.

Contrast this to the gleaming highways in the Persian Gulf's richest economies—you get the feeling they're paved with gold. And they are—black gold. Anyone who has traveled recently to Dubai, Abu Dhabi, Beijing, or the new "Asian Tigers"—Vietnam, Indonesia, and the Philippines—has stood in the shadows of a thousand cranes. In every year of late, more skyscrapers (buildings over sixty stories) have been built across the globe than during the previous year. Last year, 128 such buildings were completed, an almost 500 percent increase over those built in the year 2000. But 90 percent of those buildings were built in Asia and the Middle East. Only 6 percent—seven buildings!—are within North American boundaries.[9] Witnessing the flourishing of far-off China and the United Arab Emirates (UAE) helps you understand in your marrow that these foreign booms in construction and development must be what it was like for us—for men like my father and women like Gertrude Comfort Morrow—when we designed and put up the Empire State Building, the Golden Gate Bridge, or the Grand Coulee Dam.

Sure, there's the rare place like New York City that never rests—always new buildings going up, always old ones coming down. But there are still places in New York, and in many other once great American cities, like Detroit, Michigan, and Gary, Indiana, where you get a terrible feeling of loss lurking over the derelict skyline, the boarded-up stores, and the abandoned homes. It feels almost as though our time has passed—a terrifying and sad thought—and we're gagging on the dust left by those people and places overseas that have overtaken us. We cannot allow that pessimism to creep any further into our country. When we're knocked down, we Americans don't stay down—we rise.

Back to my bumpy ride through the city. A couple of miles from the Queens Midtown Tunnel, a guy in a BMW cut off a cable truck, and I was not at all surprised when the truck driver expressed his displeasure with a highway salute consisting of a fist and certain finger—"the bird." That's an example of the unfortunate rudeness and lack of consideration for each other that we've gotten regrettably used to. It now appears to have gotten the presidential seal of approval. You know what, though? It doesn't provide a relief valve for our stress; it only piles it up all around us like so much concrete.

What happened next, though, floored me. These guys both rolled down their windows and launched tirades at each other. It started with the usual, "Who taught you how to drive—Helen Keller?" Then it rapidly degenerated to barbs about the other's perceived politics. And finally, of course, to race. It was a little microcosm of what's happening in America today.

When the fight calmed down, I sat staring at 1 WTC in the hazy distance. Given our differences, how the hell did we ever pull that off? Despite the metaphorical and occasionally literal fingers we flipped each other—despite the sometimes knock-down, drag-out fights over nearly every hairsplitting aspect of that building—we somehow managed to work together to make a phoenix rise out of the ashes of the worst modern tragedy we Americans have suffered. As Governor Pataki says, the tallest structure in the Western Hemisphere is no accident. All around this great city there are earlier monuments that were erected against impossible odds and built at enormous sacrifice. What do these projects have in common? What lessons can we use to re-engineer a country left to crumble?

It was my job to re-engineer Lower Manhattan after the devastation in September 2001. It couldn't have been more arduous or time-consuming, more maddening and at times inefficient. And it couldn't have been more imperative.

Today, we have to face the facts of our collapsing physical environment just like we did that day when we saw massive wreckage of the buildings, streets, sewers, tunnels, subways, electric grid, and other utilities. Business, industry, livelihoods, health, psychological well-being, safety and security, travel and tourism, transportation, storage, waste, entertainment, public services, religion, the entire micro- and macro-economy—all of that and much more was on the line in September 2001, in the nucleus of the nation's most populous and commercially critical city.

Today, we see the global "electrons" that revolve around that nucleus—everything from the collapse of the Swiss national airline to a dramatic increase in fungi and crop-wrecking insects in our heartland. It's all connected in an intricate web. Burying our heads in the rubble like an ostrich should never have been an option. Yet we do that daily rather than confront the fragments of an inoperative America.

When I talk about the deteriorating physical infrastructure in America, I think of its separate parts—the crumbling roads, the aging sewers, the failing bridges, the polluted waterways, and all their less perceptible correlations in our national health—as being like little viral "deconstruction workers." The integrity of each part depends upon all the others, and each affects the whole.

With the right kind of force in just the wrong place—or with enough growth—comes the inevitable tipping point. The whole interrelated system can come toppling down. Remember Katrina. In a couple of hours, a predictable natural disaster totally crippled a major city. But what happened in the next few *days* provides a stark warning about how close we live to the edge of complete societal breakdown.

Let's for the moment forget external terrorists, terrible threat though they might be. Some of the greatest potential dangers we might soon face are right beneath our feet. All we need to do to ensure that the maximum disaster befalls us is to sit around and do nothing at all. Let's not wait for the emergency but tackle the crisis while we have some semblance of a grip. Now is the time to fix what we know is broken. If America is capable of successfully building the kinds of pipelines, buildings, and neighborhoods I had the privilege to work on, it is certainly capable of keeping its infrastructure safe and functional. Where there's a will, there's a way. We just need to vocalize the will from the ground up and get our leaders to hear it and act on it from the top down.

There are hundreds of thousands of workers trained to do the hard work of rebuilding and improving these resources—they simply need marching orders and the right leadership to make it work. Imagine what kinds of marvels would be possible today with all the latest technologies at their disposal! We know where practically all the hitches and glitches lie—the government pays engineers to assess and reassess our infrastructure problems all the time—so all we're missing is the collective will and leadership to do anything about it.

Partisan Rancor and Infighting

We are a house divided. And you know what Lincoln said about that: we cannot stand it much longer. We relish fleeting moments of unity, glimmers of what could be—or what we nostalgically think we remember—on rare occasions. Occasions such as today when I write this, after a nutjob shot up a congressional baseball training practice for a charity game.

The events of September 11, 2001 truly devastated us, but on the other side of the coin, that tragedy—like others, such as Pearl Harbor—brought us together. Yes, it might have been short-lived, but it was real. I'll never forget seeing George W. Bush address us through a bullhorn at the Pile. He spoke both as one of us and as our leader. I'll never forget seeing 150 lawmakers from both parties locking arms and singing "God Bless America" on the east steps of the Capitol that evening, embodying our resolve.

I don't reference either of our darkest hours glibly. We truly haven't seen our nation this polarized since the Civil War—and we've rarely experienced even the transitory concord we saw in the days after 9/11. Nowadays, voters and leaders from both wings "don't just disagree about the right way to reform health care or the true intentions of President Trump. Many despise each other, and to a degree that political scientists and pollsters say has gotten significantly worse over the last 50 years."[10] Surely that doesn't surprise anyone sentient and plugged in.

Nearly 150 years after the War Between the States we're still bearing the scars, still suffering a violent, long-simmering rift in our fundamental beliefs. The chasm grows wider every day. I have a prescription for this malady, one based not on sentimental fluff but on a resurgence of the mission and values that circumstances forced us to mobilize on a cloudless autumn morning sixteen years ago.

Greed, Graft, and Petty Politics

This rift down the middle of our nation does not really derive from specific ideological party distinctions. Sometimes it's just raw animosity and resentment. But I'd argue that, more often than not, its genesis is the root of all evil: the love of money.

Power and money are twins when it comes to politics. I recall getting just as much guff from my friend David Dinkins, Democratic mayor of New York City in the early '90s, as I got later from Republican mayor Giuliani from '94 to 2001. I'm talking about trying to spearhead neighborhood renaissances like the one Pataki and I agreed on for the dilapidated streets of once thriving inland Coney Island. Such places in the city were dangerous to life and limb, bleeding money, and hideous-looking. But some politicians refused to listen to reasonable ideas for redevelopment in partnership with the private sector. Why? An ungenerous critic would blame sweetheart deals. A more charitable criticism would point to intractability and the personalization of issues and plans. An idea was often argued not on its merits but on who would get their names on plaques.

Take the madcap concept of building a Minor League ball field in the center of Coney Island. That "improvement" did not increase the value of the surrounding area, a "shabby ghost" of its former self.[11] It didn't increase business for the straggling merchants in the otherwise abandoned neighborhood—it certainly didn't bring new business in, nor new residential developments. Not only that, but it didn't actually *serve* the local community at all. I love baseball. I'm a Dodgers fan from way back ("Dem Bums"). But that neighborhood needed real growth, youth services, and a future—not a botched resurrection of the past. It's not just that Giuliani misread the borough. It's that he hammered the project through just so he could say he brought baseball back to Brooklyn. Ha! That's the dark side of vision. It's called megalomania. We've got it in spades at 1600 Pennsylvania Avenue now.

Smart, well-meaning civic leaders can disagree as long as none of the parties has a hidden or personal agenda. So when Hillary says publicly she stands with all Americans but is really just after the minority vote, and Trump says the same but would sell civilians down the river for the sake of courting big business, there's a disconnect.

When the hidden agenda is pocket-lining, you get an even more sinister situation. I'll introduce you later to former Democratic New York state assembly speaker Sheldon Silver and real estate developer Larry Silverstein— two embarrassing examples of naked greed and self-indulgence related to the 9/11 rebuild. It's partly comical in the extreme and partly exasperating that some among this lot have attempted to malign my reputation. Ultimately, I wear it like a badge of honor that Silver called me "the most corrupt member of this administration."[12] Through hard work and a little luck, I made a fortune in the private sector before I ever began civil service—I didn't need to siphon off the government or a cash-poor city and state. That would be a stupid way to make a living.

Yet old-fashioned graft is alive and kicking in the halls of the regime. It is nothing new, of course. For senior officials, bribes and "gratuities" in the form of free stuff have been public knowledge since at least the Grant administration. Of course that all started way back with shiny rocks and other trinkets traded under the table for the tribe's new wheel contract. On a subtler level, you've got people who join administrations not because they love their country, but to beef up their résumés in order to eventually quit the government job and rake in the green as consultants or lobbyists.

I've butted heads with almost every major New York state and U.S. leader since the early 1980s—that's bound to happen. I've been called "temperamental," a tastemaker, a "rainmaker,"[13] a "classy political opportunist,"[14] a "bagman,"[15] a "colorful, sometimes controversial, immensely successful chief fund-raiser,"[16] and some more epithets. But I never backed down from my moral principles. For this reason, even most of those who have disagreed with my foresight or approaches to problem-solving ultimately came to respect me and invited me back to the table time and again. To make a point, Pace University, upon awarding me an honorary doctorate in commercial science, praised me for taking the lead in energizing economic recovery and overseeing federally subsidized programs for New York after 9/11, declaring, "[Y]ou, Charles A. Gargano, are a man who can take ideas and find a way to make them real."[17]

The thing is, some people saw it as a competition. I had the ear of the governor and a senator or two. I had my finger in a lot of important pies. Some people felt threatened by that. It's true there was *potential* for impropriety—I raised funds then worked for administrations through which I funded others' projects. We had strict ethical and legal ways to avoid financial misconduct. For example, I had already sold all my shares in the general contracting company Posillico by 1986. Still, Republican strategist and former Trump cheerleader Roger Stone has roundly savaged me over the years, even as the *Weekly Standard* called *him* a "boastful black prince of Republican sleaze."[18] Manhattan district attorney Robert Morgenthau, on the other hand, unsuccessfully investigated me—one of many aborted attempts to smear my reputation and diminish my influence. There was nothing there. Never. Which brings me to...

Government Waste and Inefficiency

How can you blame anyone for giving up on government-run systems and services when, if it were a public, commercial enterprise, it would be bankrupt or put out of business after a nepotism scandal? And if it was an individual, it would be a debtor's prison cautionary tale. Here, for all the "Yes We Can!" hopey-changey gimmicks of the 2008 election, all the president's men and a gaggle of oil-industry bigwigs couldn't even plug a damn hole in the ocean that sprayed 210 million gallons of crude for eighty-seven days. Both parties are guilty of ineptitude mingled with cronyism, famously symbolized in Bush 43's "Brownie, you're doin' a heckuva job" statement—not only tone-deaf but empirically untrue—while the Gulf Coast was devolving into *Lord of the Flies*. When you open the honey jar, you'd better expect the ants. Our civil servants need to serve—and not just their own schemes and press. Trump's take on this is one of the many things that causes me anxiety in the middle of the night.

Can we really "make America great again" while our president is sowing division and tweeting every petty peccadillo, even as he installed his entire family in critical positions inside a bungling bubble?

There's so much wasted potential in Washington; there are so many times when the two parties decide to score political points instead of serving the American people. Too much division and not enough of the right addition and subtraction. Our leaders actively obstruct and thwart progress on partisan grounds, with both sides blaming each other, of course, for nothing ever getting done except for the occasional "bridge to nowhere." Educated Americans overwhelmingly despise their own Congress, with a whopping 66 percent holding a negative view (and only 7 percent rating the institution as good or excellent).[19]

Spending is obscene. I'd estimate that our government wastes about $300 billion of taxpayer money per year. This happens in a number of ways. First, *overlap, redundancy, and fragmentation*. Too many official factions provide the same or similar services. Guess how many regulatory agencies, review boards, commissions, committees, geographic representatives, political groups, supervisory authorities, and organizations (victims', environmental, engineering, religious, archaeological, legal, financial, and the like) were involved in the rebuilding of Ground Zero. I don't even know the answer. A lot. It was an alphabet soup of sometimes duplicative, often contradictory, and occasionally outright inimical authorities—and we're not even talking about the fringes and the über-factional cliques. The neighborhood. The area. The borough. The city. The state. The state across the river. The Feds.

Of course everyone wanted a piece of the project; everyone wanted a say. And in many cases, these groups provided key oversight. People with knowledge and experience had to oversee the collection of artifacts and objects that would be of historical or legal significance. The Port Authority charged a team with combing through the wreckage daily at Ground Zero and Port of Newark warehouses and scrapyards to preserve artifacts that might be of historical or other value. Another specialist agency had to make sure we exceeded fire codes. Still another had to approve our wastewater engineering. And so on. What was I supposed to do with a Revolutionary warship that workers dug up twenty-two feet under the site? I get that. But there were so many competing voices that it often amazes me that our hundred-plus contractors were able to install a single orange electrical cap, let alone build a supertower of epic complexity.

The controlling governmental or quasi-governmental parties with their fingerprints on the Ground Zero rebuild show us in miniature what it's like to get anything done in America. We face self-perpetuating overregulation—reams of rules.

Just look at our tax code. Millions of miles of the reddest red tape. Hoop after hoop to jump through to get a project approved, underway, and across the finish line. Hoarding of intel—zero interagency communication or deliberate obfuscation. At the least, the kind of muddled exchanges—different systems, acronyms, codes, and procedures—among the various first responders and military/civilian aviation authorities on September 11. We learned from our failings on that score—at least we should have learned.

We need *less* government, not more, to streamline the process. Only the most essential and only those who have proven themselves effective and efficient. And everything needs to be simplified and made sleek. Plain language.

Another facet of governmental waste is *project overruns*. You say you can build it in a year? See you in five. This happened, infamously, in the Ground Zero rebuild. Let me concede here that sometimes, for the right reasons, you need to spend time—to invest it—to have any hope of a big payoff. I'm not talking about that kind of calculus. We weren't going to pull off the re-engineering of a colossal neighborhood over-night. I'm talking about incompetence, such as $100 billion a year in improper payments (in one recent ten-year span, federal agencies doled out about $688 billion to the wrong people).[20] There's no excuse for that. That alone would almost pay for our entire federal infrastructure re-en-gineering project.

Instead, we keep making the same "mistakes" year after year. They cease to be mistakes once we've identified them and still repeat them—then they're deliberate *decisions*, and they're the definition of insanity. But what's the difference when we can literally just print more money? That shows the maturity and economic reasoning of a third grader. Our president sometimes displays a similar temperament, but we need to count on his decades of business experience to re-engineer our economy, starting with radical changes to the federal budget. Am I hopeful he can pull it off? About as hopeful that he'll eventually get around to his infrastructure dream. In other words, it's possible—but I'm not holding my breath.

Some of these factors work in clusters that swell like a cancer, infect-ing critical projects. The symptom is government waste, but it's not the disease. Pride causes delays. Ego infuses stupid competition, which leads to overruns. Dishonesty and ineptitude go hand in hand—that's how you get caught. Just look at New Jersey Republican Governor Chris Christie for a view of this kind of ugly tumor.

A little background: The Port Authority of New York and New Jersey supervises critical infrastructure within the interstate region's trade and transportation corridor—a 1,500-square-mile port district that encompasses twenty-five miles from the Statue of Liberty in every direction. That includes five of the region's airports, the New York/New Jersey seaport, the Port Authority Trans-Hudson (PATH) rail transit system, six tunnels and bridges, the Port Authority Bus Terminal and George Washington Bridge Bus Station—and the WTC site. The Port

Authority operates its own 1,600-member police department. It serves seventeen million people in the greater New York City region, where nearly nine million people work, generating close to a trillion dollars a year in gross regional product.[21] Of 196 countries in the world, only fifteen belong to that $1 trillion club in terms of GDP. The Port Authority region is a monster economy—and monstrously hard to manage and maintain.

The agency, which I helped run for twelve years, was originally formed in 1921 by an Act of Congress, some say to lessen parochialism and infighting between the two sister states. Really, Congress just wanted to stimulate the economy of this vital part of the Northeast, and build the necessary bridges, tunnels, roads, and airports serving the New York City area. But ever since its noble beginnings, there has been constant wrangling between the two states.

Fast-forward to 1987, when New Jersey execs claimed ownership of the Statue of Liberty. Please. The always amusing Democratic mayor Ed Koch held our ground and famously remarked that she would "stand exactly where she has stood for one hundred years…facing us and showing another side of her personality to New Jersey."[22]

Decades before that, the two states battled over commuter taxes, and a decade after it, my boss, Pataki, and his counterpart, Republican New Jersey Governor Christine Todd Whitman, stood off in a wrenching mêlée over Whitman's ego about Pataki's leaving her out of the loop on a management decision he made—a decision well within his rights and responsibilities by long-standing precedent. The fight really knocked the wind out of the Port Authority. Later, Democratic New Jersey Governor Jon Corzine was petulant about New York City Mayor Michael Bloomberg's congestion-pricing plan, and cannons were launched from both sides of the Hudson. It's a boring tradition at this point, and a huge waste.

Here's the thing. New York and New Jersey are not equal in any way. New York state's population in 2017 was a little under twenty million, making it the fourth most populous state after California, Texas, and Florida. New Jersey's population in 2017 was less than half that—fewer than nine million people.[23] The entire state has about the same number of people as New York City alone[24]—population 8,550,405 in 2017.

At the same time, the economy of New Jersey is worse than that of New York. New Jersey's gross domestic product (the sum value from all industries in the state) in 2016 was $581 billion. By contrast, New York's was $1,487,998,000,000—that's $1.5 trillion, nearly three times higher.[25]

In fact, it's that latter problem that has caused the most interstate stress inside the Port Authority. As a rule, New Jersey depends much more on the Port Authority and its ability to build and maintain things. They demand that they need more of a share. It's true they've been hit harder by downturns in the economy, by unemployment and lack of opportunities. For residents of each state struggling on the lower income level, it's definitely harder to get by in New Jersey, where the hourly minimum wage in 2017 was $8.44, contrasted to New York's, which was $9.70. New York City's hourly minimum wage will go to fifteen dollars by 2020[26]—a plan I support for the most part, as it has worked wonders (though it's not an economic silver bullet) in places like Seattle, despite some vocal detractors.[27]

But none of that excuses the actions of the most unpopular governor in New Jersey's history, the "I-don't-care-about-optics"[28] Chris Christie, aka "Jesus Christie." I'll give you one example—the Access to the Region's Core (ARC) train tunnel. I knew that project very well because I was approached about it as the head of the Port Authority of New York. New York Republican Congressman Jerry Nadler used to come to my office a lot, and we met about the Secaucus-New York City rail link several times. I agreed with him that it was a good idea, and was necessary to increase Hudson River travel capacity.

The ARC tunnel project would have included lots of new infrastructure, such as tracks, a new rail yard, and a tunnel under the Hudson River. A splendid new station next to Penn Station would have been built to accommodate more trains than can currently be handled as long as the useless Amtrak continues to "operate" it. It would have replaced an extant 110-year-old train tunnel whose overhead wiring causes nonstop delays and problems and poses a real danger to commuters. We estimated a project cost of $8.7 billion.

With all parties on board, we started construction in 2009—it would have been completed by 2018. Then in 2010, Christie nixed it. Stopped it in its tracks. He said he was worried about cost overruns and claimed his state was too cash-poor to continue. It's true—Jersey's a mess, thanks to bad leadership for a long, long time. Problem was, we'd already spent heaps of money on ARC. They say $600 million[29]—I think it was closer to a billion. Really, Christie just thought New York wasn't paying its fair share. So he punished his own people and further eroded his already-in-the-tank economy. Then he snatched Port Authority funds previously earmarked for the project and repaired local New Jersey roads instead.[30] Not cool, Christie. I agree in principle that it's usually a better idea to fix and maintain existing infrastructure than to build spanking-new pet projects that seem cool. In the case of ARC, though, I was convinced it was a good idea, if only because of the indecent amount of time, money, and manpower we flush down the toilet between the two existing tunnels and the George Washington Bridge because of delays. We need a new means of crossing that river. It's a lifeline for both our economies.

We could have worked together on that. Instead, New Jersey dipped into the coffers of the Port Authority in a noncooperative way. They basically stole from us.

For the record, New Jersey doesn't have a monopoly on terrible leadership. Democratic New York Governor Andrew Cuomo is not the man his father was. He's almost as bad as Christie in terms of wasting the state's money for the sake of image.

Even when you don't have a corrupt or foolish leader, the way budgets traditionally work still needs an overhaul. I'd start with rejiggering all those "use-it-or-lose-it" budget lines. Nearly all government agencies, with the exception of the Department of Defense (DOD), don't get rewarded for saving money. If they don't spend the budget they requested, that money goes back into the general fund—they lose it. So you see a lot of last-minute "Oh-we-really-need-X-as-soon-as-possible" requests right at the end of the fiscal year. If that's not the way your home budget works, it shouldn't be the way we do business as a nation.

And how absurd are our federal procurement laws? You remember the DOD's paying thirty-seven dollars for each screw, $7,622 for each coffeemaker, $640 for each toilet seat, and a whopping $2,043 per nut in the mid-'80s?[31] That insanity has been shuffled into nonmilitary contractor fees. The General Services Administration (GSA) publishes a calculator for potential government contractors, providing ceilings for given services. Did you know that a GED and a mere four years of "experience" qualifies you to be a senior engineer consultant at a rate of $226 per hour?[32] Nice work if you can get it. If transcription services can be billed at hundreds of dollars an hour,[33] then I picked the wrong profession! Even clerical workers with high school diplomas and minimal experience can command nearly one hundred dollars per hour on government contracts, and a middle-level manager can cost the taxpayer double that amount—at least three times the going rate in the private sector.[34]

The funny thing about those rates is that they're *published*. The government acknowledges its own wastefulness. In most other areas of administration, it's lack of transparency that's the problem. For all the campaign pledges and eloquent speeches, the Obama administration was more secretive than Nixon's. For example, did you know that in March 2015, Obama's crew quietly voided a regulation requiring the administration to comply with Freedom of Information Act (FOIA) requests—a liberal darling if ever there was one—thus exempting itself from public scrutiny and oversight?[35] I know I said we could use fewer rules, not more, but we need to keep the necessary ones. Our ability to check power is essential to the fair running of our enormous nation. For all the terrible ways the press has sometimes treated me and the agencies I've served, I would never call our fourth estate "the enemy of the people."[36]

To be clear, even as big projects take time, sometimes they also take money. Fiscal responsibility doesn't mean curbing spending per se. It means spending sensibly. What we need to curb is *waste*. It's the number-one way we can begin to pay the enormous sums necessary to re-engineer the whole U.S. infrastructure and serve the public in a way that will spur our growth.

Entitlements

Government exceeds its authority all the time. This is because it exaggerates its worth. And that, in turn, derives from the misapprehensions about its role, its purpose, and the best way to conduct its affairs. Which is, ideally, *hands off*. There's a difference between leadership and mere management. Any manager can put a ladder up. It takes a leader to know *where* the ladder should go. It's the difference between serving and infantilizing. Government overreach into every aspect of our lives—even it were effective—breeds a deep dependency. Government should *give care to* people when necessary and where appropriate—not *take care of* them under any and all conditions from womb to tomb. The accident of our birth as Americans in the twenty-first century instead of as peasant Mongolians a thousand years ago does not entitle us to tip-to-toe platinum-plan coverage in perpetuity. The Constitution is not a landscaping contract your grandparents fortuitously purchased from a cemetery before you were born. Our citizenship is a privilege and a blessing, and, as such, must be earned according to the maximum of our individual abilities. Taking our liberty for granted while at the same time holding our hands out—what's wrong with that picture?

The inevitable feedback loop of this process, the parental model of government, keeps people childlike. Yes, many so-called entitlement programs make sense. Along with the ten thousand-plus who turn sixty-five every day in America, I contentedly cash my Social Security check after decades of contributions. But for two-thirds or more of Americans, that monthly check represents half their monthly income or more.[37] That was never its intention, and rightly so.

The *expectation* that the government will take total care of all our needs, cradle to the grave, is nothing like what the founding fathers intended. We had a king once—and shed oceans of blood to stop being patronized. This country was built on self-reliance, self-sacrifice, self-esteem. In what esteem can we hold ourselves when we need the state to change our diapers for us? We need to gradually wean people off welfare, food stamps, and other handouts. We need unemployment and disability safety nets, of course, but we must take more effective steps at curtailing the manifold abuses we know are happening. This is a cultural problem we need to tackle from the ground up. Even the ongoing dispute over health insurance—is it really a right in the proper sense of that term?—comes down to the difference between a kind of enslavement and empowerment.

I'm not suggesting we suddenly pull the carpet of benefits out from under the neediest of our brethren—infants, the infirm, and so on. No one should feel shut out by the government. But its job should be stabilizing people, providing opportunities, balancing inequality. We can help American citizens' thought processes about such expectations evolve more toward self-reliance, discipline, and hard work—that's a good role for government. We first cut the grossest misuses while simultaneously streamlining the administrative bureaucracy. Then we endow career counseling (job training, retraining, and placement services); early education; community college; child development; crime prevention and criminal rehabilitation; neighborhood and urban renewal; public-private partnerships; emergency services; universal internet access; and other incentives for the kind of self-reliance that founded our country. It would save money now and in perpetuity because we'd stop proliferating an exponentially more expensive system. The overall benefits as related to costs would be almost incalculable. We can keep gift-wrapping that money and giving it away or we can *invest* in the future of America. We can keep doling out pricey gruel to compensate for our past failings or we can teach our citizens how to fish better.

Unchecked Union Power

To paraphrase *Animal Farm*, "private unions good, public unions bad." Public unions are inefficiency experts—geniuses, in fact. Unions in business have their own problems—don't get me wrong. But over the past sixty-plus years, I've found that their members are almost always well-trained and educated in their field. They—plumbers, electricians, welders, and the like—set a high standard and, in my experience, they're always prepared to undertake hard work and do it right when it really counts. Union workers bring a great deal of general knowledge and task-specific proficiency to every construction job. Union steamfitters know their jobs inside and out, but probably couldn't put up Sheetrock any better than the average orthodontist. A steamfitter doesn't have to. He's a master steamfitter. To that guy, steamfitting is the most important part of the overall job. That's as it should be for the drywall crew, too, and the electricians, and so on.

All told, there were forty-nine unions whose thousands of members got their hands dirty, showed up in all kinds of weather, and risked their lives at Ground Zero until the finishing touches were done on 1 WTC in 2014.

Granted, in some cases, even with the qualified and extremely skilled union workers, you do end up with a few crew members who are mostly useless, or do far less than the other workers. You have no choice but to bring them on because of union overregulations. Contracts stipulate certain minimums for how many people you must have on a given operation. For example, on a sewer job, if you have three pumps within five hundred feet, you need an engineer on every pump even though one engineer could easily cover the three. You might be able to get away with a four-person steelworking crew on a tower build—one to tie the steel, one to bring it over with the crane, and one or two to line it up. Collective bargaining might mandate a couple of extraneous guys or gals. It's a trade-off. Maybe getting the best people with superb skills might be worth sacrificing a bit more money, especially where safety is concerned. Or perhaps together, workers and management ought to look at negotiations for new union contracts together—after all, it is called *collective* bargaining—with all eyes on maximizing efficiency rather than padding.

In any case, private unions come with so much less abuse than public employees' unions. They find more ways to cut corners and take advantage of management than you can even imagine. Do you know how many Long Island Railroad (LIRR) workers have gotten approved for disability payments by the time they reach retirement age? It's a mind-boggling 96 percent of applicants.[38] Can you imagine the gall it takes to do that? LIRR may not be world-class transportation, but permanent injury to hundreds of workers is certainly not what's happening on their tracks and in their stations. Their union leadership claims not to know about the practice, but how can that happen without the organization's knowing anything about it?

It's no different in our public schools. The Yonkers (New York) Federation of Teachers was caught gaming the system in a journalist's sting, during which union leaders advised a member to lie about abusing students and to break rules to avoid being punished.[39] New York City is no better. In 2009, there were about six hundred idle teachers sent to one of six large temporary reassignment centers, commonly called "rubber rooms." There, the failed teachers played cards, napped, listened to music, left for lunch breaks, took summers off—did squat, really. There they waited happily for a year or more for disciplinary hearings on accusations that they were incompetent or had hit or molested students. They received full salary and benefits while they waited.[40]

The rubber rooms were supposedly closed in 2010, but the deal to shut them down is routinely violated. According to some estimates, there are still two hundred to four hundred city teachers reporting to "work" each day in one of the rubber rooms, at a cost of fifteen million to twenty million a year in taxpayer dollars. Most of those teachers "earn" at least $100,000 annually for their idleness.[41] They give good teachers like most of the ones I had a bad rap.

The fact of the matter is, union leadership encourages public employees to take as much as they can get away with at every level of government, and this erodes trust among the very people funding them. That kind of abuse and inefficiency doesn't translate into greater productivity or service to the public.

Before serving George Bush Sr. in the State Department, I had the honor of working directly with Ronald Reagan. My trial by fire was getting appointed at a critical juncture, as deputy administrator of the Federal Urban Mass Transportation Administration, also known as UMTA (part of the DOT; later renamed the Federal Transportation Authority [FTA]).

This was August '81—a date that will go down in infamy in the public union annals. Reagan had been in office half a year. I was ensconced as a partner in a big construction firm I'd grown on Long Island. Now I found myself in Washington, meeting with Drew Lewis, the secretary of transportation, and about ten other people, regarding the president's offering me the position at UMTA. The phone rang. Lewis pardoned himself. "You'll all have to excuse me. I just got a call from the White House. I have to get over there." I listened as other assistant secretaries buzzed about the call. One of them said to me, "Know what that's about?"

I shook my head—*no idea.*

"The Gipper's firing the air traffic controllers today, despite the [transportation] secretary's best efforts to avoid it."

Reagan hated the union's stranglehold, especially when public safety was on the line. I thought, "Good for him." Reagan had the courage to fire those guys for endangering lives by "demonstrating" they couldn't handle everything that was going on. That was bullshit. They just wanted more hours, more overtime, more perks—the usual union menu. And the fact that they were willing to risk so many close calls in the air to make their point was despicable. So Reagan said, "That's it." *Boomp!* Fired 'em all. Laid off thirteen thousand members of the Professional Air Traffic Controllers Organization. He set back private unions by decades. I cheered. I accepted the job.

Fifteen years later when Pataki first appointed me as his chief economic adviser, I used Reagan's leadership as a model of expanding efficiency in the agencies I oversaw.

Bad Leadership at the Top

I'm not the kind of guy who bashes the American president, but these are extraordinary circumstances. I think one of the several reasons our extreme (to say the least) president managed to pull off the remarkable and surprising upset he did in 2016 is that so many Americans feel like they've lost basic control of their own lives, their private spheres.

There are two factors interoperating here. On the one hand, a populace feeling out of control is a populace asking for a strongman—you get the government you deserve. You know the old idea that kids—especially wild ones—are all looking for discipline, a firm hand? It's the same with us adult citizens. So let's say you don't feel like you've got a handle on your finances, your family, your future. Along comes an alpha male beating his chest and promising the world—security, stability, "law and order." How's that not appealing?

If you really look at this, though, you'll see the deeper dynamic going on, which strikes me as quintessentially anti-American: you might be poor, your family might be disintegrating, your job prospects dim—but at least you're not Mexican, right? You're part of the winning "us" instead of the "loser" other. This is the dark side of American exceptionalism, which sometimes creeps into our conscious or subconscious attitudes and behaviors—especially when we see it modeled daily from the top.

Yes, let's absolutely make America great again. But clinging to a xenophobic interpretation of that sentiment isn't going to forward anyone's practical prospects on a personal or national level. Don't fall for that desperate—not to mention insincere and grasping—last gasp of a culture on the wane, clinging tenaciously to the dregs of its former glory. That's sad. That's not where we are, have ever been, or are going. It's not what we're about. This is not the end of the Roman Empire, all of us partying while the barbarians bang at our gate. Our real glory lies not in our vain boasts of greatness, but rather in our history of ingenuity, our humanity, our fierce work ethic, and the astounding, unparalleled realization of our particular strand of democracy. If we expect to lead, we must do so not by conceited words but by considered, careful example.

Are we putting forth the best example we can right now? I don't think so. At the same time as we're enduring humiliation abroad, few Americans' stations are rising at home. It's the exact opposite of the brief world respect we earned by our resolve in the wake of 9/11 through our earnestness, our humility tempered by grit and steadfastness. Remember that? It was before Obama's apology tour, for which he somehow won a Nobel Peace Prize in 2009, which shocked him and everyone else, and which even the Nobel secretary soon regretted.[42]

Of course it's ridiculous to expect the U.S. president will or even *can* actually return anything you feel you've lost in your life. It's not like anyone who voted for this president expects him to lift the lid and put a chicken into their personal pot. We're not a stupid people. That's not why Trump is the supposed leader of the free world. It's because in the face of our hopelessness, he wheedled hope. Part of our real greatness is the opportunity I have to speak up about what's not working, as well as what is working, without fearing that the "thought police" will roust me in the middle of the night to shackle and torture me. That does not mean all's well that ends well or let's ride it out and hope for the best because at least we're free to dissent.

No, we must do more.

The rise of Trumpism saddens me not so much for political reasons, but because it's emblematic of a general sense of incivility, contempt, and entitlement to intimidate and isolate, rather than to bolster and include, to put down instead of lift up. It leaves people behind and further reduces us. It's not about service but about polishing another Trump trophy. Not to mention it isn't Christian or charitable by any standard.

Yes, we have real problems. And the Left is certainly responsible for much of it, if not most of it. But it's the president's responsibility, his political and moral obligation, to unify us and to represent the best of our traditions and the core of our values. Instead, this guy's acting like the class bully, the stereotypical boorish and "ugly" American. It's embarrassing. And there's nothing either Republican nor even truly conservative about the behavior or the policies—if you can call them such—behind them. Trump has no intention of making America great again. He just wants to make Trump even better in his own eyes.

But the fact such a man could become president when so many other, worthier candidates vied for the part in the primaries means that for many of us, such behavior is okay now. Anything might have been at least marginally better than the turmoil we would have suffered under Hillary, but that doesn't mean we should tolerate loutishness in our highest office. Graciousness and the core Conservative values of compassion and altruism have clearly eroded so far in this country, I'm afraid it will be very difficult to restore them. Beyond that, our president has assaulted objective truths with several knockout punches.

I've had many dealings with Donald Trump. And he has spoken highly of me. He's a charmer, a personality as big as life, and a clever businessman, too. But I've found in all those dealings that only one person in the room (or the hotel, the plane, the White House) really mattered to Donald Trump—and that person was Donald Trump. Even his argument that family ought to come first flies in the face of his real duty—to put country first. Leaders like Lincoln and Washington sacrificed themselves and their own exaltation for the greater good. I'm certain Donald Trump, for all his talk of loyalty, always thinks of himself and puts himself first. He always wants to come out on top.

Where does that leave the American people?

Trump is used to dealing with the private sector as a successful private citizen and CEO of a private company. He needs to take a civics class to learn what it means to be president. This includes mastering the art of democracy and understanding how it differs from his dictatorial style.

Let me reiterate that I was no fan of Obama's, either. His foreign policy, especially, left a lot to be desired. And he had the opposite problem of Trump most of the time—he couldn't pull the trigger on anything for fear of how it might smear his sterling reputation as a darling of his staunchest supporters. These are totally different but equally disturbing deficits. Surely we can do better than these two guys.

Terrorism

Speaking of foreign policy failures, our actions—despite some good efforts and questionable judgments alike—have contributed to the expansion of radical Islam. At least we have not nipped it in the toxic bud it was on the day in 2001 when nineteen terrorists from the Middle East took down the center of our trade with our own commerce.

Those insane elements are closer to their caliphate cause than ever, and further than ever from the sociopolitical underpinnings that we sent "on the march" in an ill-advised attempt to insinuate ourselves into a millenniums-old ancient antagonism we could not possibly grasp. We toppled some terrible tyrants, yes, but at what cost?

It's only our geographic isolation, really—the oceans that buffer us—that have spared us the fate of our European allies. Their streets are under attack by guerillas hell-bent on dying to teach us some lesson that I—and most of the West, I suspect—cannot fathom. Slaughtering babies strolling with their mothers in France and England and Germany? Blowing up airports in Belgium? Beheading kidnapped Western soldiers and journalists? Martyring themselves by taking out as many of us infidels as they can sweep with their stolen trucks? That's going to happen between our shores again unless we radically re-engineer our policy and our entire thinking about the problem.

Meanwhile, domestic terrorists born of the radical Left before I was born, and who enjoyed a heyday in the 1970s (think Symbionese Liberation Army, the Weather Underground, the Black Panthers) still threaten to tear us apart from the inside. In the wake of dramatic foreign attacks from Pearl Harbor to 9/11, we forget the galling and foul homegrown kind. Oklahoma City. Centennial Park. The Unabomber. The JDL. The Boston Marathon. San Bernardino. Fort Hood. The evil anti-gay massacre in Orlando. Dylann Roof and his white-supremacist ilk. Various anti-abortion bombings.

I'd rather not even mention the proliferation of school shootings since Columbine, the most disgusting of which happened in Sandy Hook, Connecticut, depriving twenty families of their precious little children.

We must find ways to ensure the benefits of our unique brand of freedom don't include the freedom to choose murder to ameliorate a sense of individual disenfranchisement. That particular issue has nothing to do with access to weapons, by the way. It has to do with people, mostly young, but sometimes too old for that shit, who don't feel respected. They seek respect, paradoxically, by committing the most disrespectful, despicable crimes against innocent bystanders. Then they rage and express self-disappointment.

Misguided Protesting

That's why so much of the unproductive protesting I see worries me and angers me so much. I don't understand what these people are thinking. Isn't it better to *participate* than vainly protest? Protesting is impotent in contrast to proactive action. You could spend all day complaining about things you don't like, or standing ineffectually in a subway station with an obnoxious sign, but that doesn't change a thing. Working to make a difference is what changes society—and if you love this country and the people in your community, you need to work for them, to be a part of the solution.

Don't misunderstand me. I'm not saying, "My country, right or wrong." I'm not saying, "Love it or leave it." But effective and necessary protest movements like the Civil Rights actions of the '60s—especially the ones based on the passive-resistance idea championed by Mahatma Gandhi and the Reverend Doctor Martin Luther King Jr.—had a clear mission. For the most part, participants were trained and understood the civil limits to their actions. They aimed to educate and encourage change without trying to undermine the very country that allowed them the right to be heard without fear of further persecution, imprisonment, excommunication, and execution. In other words, they weren't anti-American.

I noticed a departure from that kind of peaceful and productive protest around the time of the Vietnam War demonstrations. The groups demonstrating and "occupying" parks and highway on-ramps nowadays remind me of that. Masked loons hurling Molotov cocktails through bank windows and chasing after the limousines of G-20 leaders, crying with two loaves of bread under each arm. Children of relative privilege "occupying" Wall Street while sipping their overpriced caramel macchiatos from Starbucks. It makes no sense.

It's only a matter of time before that devolves into chaos. So you strongly support the right to life? Your way of showing that is to bomb an abortion clinic and kill a poor, young mother and her doctor? You're protesting a perception that the police have singled out your neighborhood and your people? The courts have abandoned your cause and will always side with "the Man?" Why not riot, loot your neighbors' stores, flip and burn all the cars and buses those same people depend on for work? You demand the police protect and serve you, so your solution to that is to march with signs saying "Pigs in a Blanket"? And to take potshots at officers risking their lives doing exactly what you demand? If black lives matter so much, why the hell are you damaging your own black neighborhoods? If you're supposedly defending the free speech rights you learned about in college, why mutiny on your own damn campus, turning the whole thing upside down?

I'll tell you why. None of those demonstrations is really an organic, spontaneous, true expression of woe with the aim of mitigating it. No. Most, like the recent protestors in Baltimore, are groups of pawns whipped up and encouraged by we-all-know-what agitator elements. Insert the name of your local Al Sharpton or Louis Farrakhan here.

Erosion of Conservative Values

This all comes down to that one key thing. The beliefs and practices we founded our country upon, and which sustained us through our worst days and nights, matter not to the younger generations. We're not united. Not like we used to be. We're not hardworking like we used to be. We're not as responsible. We don't value the traditional family as we had since time immemorial. We don't respect our rivals anymore, but instead denigrate them to the bone. We've ceased listening to each other and compromising in healthy and fruitful ways when necessary. Though we've always been a nation of immigrants, today's average newcomers don't assimilate like their ancestors, like my grandfather did, *because they want to be American.*

So our survey shows us a pretty grim and overwhelming picture, right? Well, mark my words—America will rise again. But not if we don't all show up and punch in for the job ahead. That's not all we need, though. The best tools, hard hat, and work ethic won't supernaturally erect a supertower. There are several critical steps in between. First we have some cleaning up to do.

2

Remediate and Personalize

You First

Re-engineering America starts at home, with our own lives. Think local and long-term. Let's start with local. When I was a kid in Mrs. Lambert's fourth-grade class, there was a sign on the wall that said "Keep Your Area Clean." I assume they still teach kids this valuable lesson. But it seems that many of us forget that wise advice in adulthood. Personal responsibility is just that—personal. You build it from the ground up. Even if you think global, you have to start local. You have to build yourself into the person you want to be, based on your values, which we'll talk about continually throughout this book. Find yourself a role model or mentor out there who can help keep you on track, help keep you honest. But beware. People truly worthy of that position of influence are very few and far between.

Gone are the days when good behavior was modeled from the top down. You can't expect the average leader to be a guy like Ike. If you want Eisenhower-, Reagan-, or Lincoln-like behavior, base your behavior on those paragons' actions. It sounds corny, but when I was in politics, I found myself frequently doing self-inventories, asking myself, "What would Ronald Reagan do? What would George Pataki do?" I could just as easily have asked, "What would Bill Clinton do?"—then done the opposite. No dynastic disgrace for me, thanks. In any case, I'm responsible for my own actions and behaviors above all. You are, too. The best way to change the world is one person at a time, starting with yourself.

When you find yourself bitching and lamenting about the latest scandal-addled jerk in whatever office, whatever position of power, maybe refocus your lens on your own affairs. You can't expect the Oval Office to magically clean itself if you don't begin by scrubbing your own desk. Don't expect the rest of the country to clean up its act until you straighten up your own backyard. I'm not saying you should feel powerless in the shadow of the mighty who are running amok, as they do in, say, Russia. Part of personal responsibility is participation, but before you can help sway things according to your values in the larger community, it's worth trying to gain some moral authority over your own life and affairs—to *own your own mess.*

Simon Peter famously took Jesus's advice about putting his angling skills to work fishing for men's souls. Maybe the best way to do that is not to search for exemplars but to lead by example yourself. Eleven of the twelve disciples did just that—worked nonstop to become worthy of role modeling and mentoring others. You can start that project today at home for no money down. I'm trying to say politely, "*Please clean your own Ground Zero first.*" If we all did that in our own lives, imagine the overall effect on the nation!

Focus first on the local and personal arenas, where your power is greatest, and you'll have the most potential to dramatically improve your life and that of your family. That, of course, will help all other boats rise, and will position you better to tackle wider challenges outside your own immediate circle. Think of your world as a series of concentric spheres with you in the center. Your aim is to focus your initial exertions, energy, and resources where you already possess the most influence. Some of those efforts, of course, can be toward widening those spheres. If you run for office, say, you increase you influence beyond your personal life.

Concentrating on yourself and what you can do about a greater given problem might at first strike you as depressing. You might feel helpless—what can one person do? It ought to be just the opposite. For example, let's say you really care about environmental sustainability. You have nearly 100 percent power to adjust your own consumption. If you focus all your efforts there, you've actually done about 99 percent more than anyone else is doing. Awesome. You've also demonstrated something you can informally motivate others to accomplish in their lives—or formally teach them to do. Now you start a blog based on your expertise, maybe get a few local interviews in the paper, go talk to a sixth-grade class, and otherwise work on spreading the word beyond your immediate sphere.

Let's say you utterly and completely fail to get your message across as strongly as you'd like. Only 1 percent of your message seems to be getting through to the people you're trying to reach. Okay. How many people, though? Ultimately, maybe that 1 percent is really infectious. People carry the message like a virus. A friend of your neighbor's ex-wife tells her postal carrier, and the postal carrier mentions it at a party, where five people pick up the single technique and spread it over social media. You've created a movement. A million people adjusting their energy consumption by 1 percent is a million percent, and you've just increased the value of what you alone were doing by ten thousand percent.

This focus-on-yourself suggestion does not mean do so for ego alone. You don't want the presidential historian and former Reagan speechwriter Peggy Noonan announcing in the *Wall Street Journal*[1] that senators and your own staffers have nicknamed you "Obam-*Me*," because you're so in love with your grand ideas.

Just do the thing you wish everyone would do. Do it consistently and correctly. Start privately, with no fanfare, no grand announcement. And not with the initial aim of convincing others to come over to your side. Do it for self-improvement only. Do it because it's the right thing to do.

"If we could change ourselves, the tendencies in the world would also change," writes Gandhi. "As a man changes his own nature, so does the attitude of the world change towards him.... We need not wait to see what others do."[2]

Bulldoze (or Bucket-Brigade) Your Pile

Survey your life. Assess your pile. Start your cleanup effort. Do it one bucket at a time and don't stop. Enlist help if necessary—you're likely not as alone as you might sometimes feel. Toss what you must. Recycle what you can. Know it will be costly—change will always exact a price—but keep your eye on the prize.

This might mean you have to terminate certain relationships and/or start others. You might have to end a destructive habit or two using whatever resources you've got at your disposal. It might be necessary to leave your job. To start a new one. To change your priorities, such as spending more time with your family, pursuing your artistic dream, or serving others somehow. Maybe you have to let go of old pain and resentment, forgive your parents for their human failings. It's possible you need forgiveness for some awful thing *you* did. Start by forgiving yourself, then make yourself worthier of forgiveness by the injured party, yourself, the rest of the world, and the Man Upstairs.

Then move on. If it's unclear to you whether or not to apologize outright, err on the side of doing so, and don't wait till your deathbed. Genuinely and unequivocally apologizing is not an act of weakness. It takes courage. Be clear and courageous, do what you must do to right the wrong—then move on with your life.

Living with regret and self-loathing, victimhood, resentment—all of that's a venomous mound, a brownfield. Clear it out or you can't rebuild. Clean it out or it'll kill you. If you must, board up the site, abandon it, and move somewhere else.

Start with the practical. Maybe you've suffered legitimate terror in your life, so you must do everything in your power to escape some circumstance—to flee abuse, for example. Or maybe you need only to take some night classes, finally have that talk with your boss. Stop going to so many office parties for people you don't even like—and spend that time on self-improvement or making yourself better at your job.

Most of us have accumulated a fairly gnarly pile of psychic debris. Usually, the deeper you dip into a heap like that, the more intense it gets—it's just physics. Official thermal imaging from a NASA plane above Ground Zero showed the hottest spots below the pile ranged from 800 to more than 1,300 degrees Fahrenheit.[3] But an engineering firm consulting on safety, health, and environmental factors on the site claims its helicopter recorded daily underground thermal temperatures higher than 2,800 degrees Fahrenheit[4]—which is exactly the melting point of iron. The point is, be careful you don't get burned.

Unless we can eliminate that pile, though, and keep those buckets moving, we'll never be able to build anything new, to make what we want of our lives. Take responsibility. That means no matter how dire your circumstances, you have the *ability* to *respond*. Don't be a victim. Don't wait for the dustman or the fed-up neighbors to finally hop the fence and clean up your mess because they're sick of looking at it and the rats it lures.

Just remember, whatever you do, it will not happen overnight. When it feels overwhelming, remember the cleanup and recovery efforts in that zone of annihilation northeast of Vesey and West. Workers in twenty-four-hour bucket brigades sifted through and carried out much of the 1.8 million tons of noxious debris at Ground Zero five pounds at a time—108,342 truckloads[5] of destruction. It took eight months and nineteen days to haul off, and you'd need time-lapse photography to ever see them making a dent. They just kept going, like hard-hatted Energizer Bunnies.

Foundations First

Once those guys had cleared the space, it was the only the beginning. You can't build anything—especially anything big—without a rock-solid foundation. You need to know the bottom line. In engineering, you've got to understand what you're really dealing with, and how it ended up that way.

When your goal is to re-engineer, the only way to avoid prolonging the errors of your predecessors is to assess what didn't work and discern what, if anything, is worth keeping. The good news is that you as a person—and our nation in general—have gotten a lot of stuff right. We all have a sturdy foundation to build on. They call 'em "the good ol' days" for a reason.

The country I grew up in was thriving. After shaking off the Great Depression and winning World War II, America had developed bigger, stronger, faster, more effective means to compete and prosper. We expanded on a larger scale than anywhere else on the planet. This boom animated the new middle class, building on its strengths of ingenuity, hard work, discipline, and adaptability. What my parents' generation learned from their parents before them above all else was the value of hard work. In fact, we Americans expected to work our asses off, seeking out opportunity and, if we didn't find it, making our own. No one counted on a handout. We relied on ourselves and each other (our friends, families, neighbors, civic groups, churches, and synagogues)— not on the government.

The foundation we need to bolster before we can begin any mean- ingful physical project happens in people's hearts and minds. We're all rightly fed up, disappointed, let down. Okay, we've had our whine. Now let's get the hell to work. At ground level, taking stock and illuminating our dissatisfaction with business as usual can actually spell potential for dramatic new opportunities. Somebody had to want a horseless cart, to cross above the Thames, to prove the moon wasn't made of cheese.

We need to motivate ourselves the way our ancestors did. Can you imagine a nineteenth-century farmer having time to suffer some of our modern ailments—many, if not most, invented by Big Pharma? There was no time to mope. If you didn't milk your cow at every dawn, bad things would happen. If you didn't tend your crops, your family wouldn't eat. If you didn't cut firewood, you would freeze. Who was going to do that for you? Certainly not your Congressperson. I've proposed that the government—sometimes with well-intentioned social programs meant to improve our lives—undercut us over the course of a few generations. Made us regress. Turned us into dependents. If someone puts a free lunch in front of you your whole life, why bother hunting for food? Why take responsibility for your family's well-being when the government will do it for you?

That's translated into a template for how we raise our kids, I think.

Kids Today

Of course I think my grandkids, Charley and Michael, are the best and the brightest to have ever walked the earth. But I noticed that when they were young, their expectations of their parents were startlingly different than in my generation. I got my first job delivering *Brooklyn Eagle* newspapers at age twelve. I was out every day at dawn through the shirt-sticking summers and skin-cracking winters of Brooklyn. Dogs, traffic—nothing could stop me. Within a year, I became station master, collating and handing out two thousand newspapers to all the other delivery boys—the newsies. I used to get up at three in the morning on a Sunday to insert all the comic strips and magazines into the regular paper—we didn't have robots to do that kind of task back then. I was the robot. Then I'd hand out the exact number of papers needed for every route. *Then* I did my own route.

At some point, I got enterprising, figuring out I could run "outsourcing" contracts. I paid neighborhood kids a penny a paper to run parts of my route for me, say Seventh, Eighth, Ninth, or 10th Street. I took a ten-minute breather. I read the comics: *Batman and Robin*, *Superman*, *Mickey Mouse*, and the newbies, *Archie* and Walt Kelly's *Pogo*. The war rationing of butter and sugar on the home front ended in 1947, and suddenly it rained junk food: Almond Joys, Junior Mints, Smarties, and Cheetos. I had just enough time to indulge. Then it was back to work. Afterward, I went to church and, finally, went with my father to help at his job. Most kids today are still in bed by the time I was finishing lunch.

My boss and mentor at the *Eagle*, Mr. Cavagna, noticed my hard work, and he made sure I knew that he knew it, too. I was voted the number-one *Eagle* carrier of the year, having canvassed and signed up the most new customers of any delivery boy in Brooklyn. As a reward, Mr. Cavagna invited me to the Dodgers' 1947 National League pennant celebration at Brooklyn Borough Hall. There was no greater honor—it was a waking dream—for a dyed-in-the-wool Brooklyn boy than meeting Jackie Robinson, Pete Reiser, "Pee Wee," and the rest of the gang. My God, what a night.

You know, the best part of that for me was that I *earned* it—sweating, toiling, and hustling as though my life depended on it. Through entrepreneurship. Right there, I learned the greatest of American values firsthand. I'd seen my father and mother work hard and continue to improve their station in life—and now I was doing it, too. My reward was that the players signed a baseball my brother had suggested I bring for autographs. I got Reese. Reiser. Carl Furillo, Dixie Walker, Duke Snider, Eddie Stanky, Gil Hodges. And the great man who'd busted the color barrier, Jackie Robinson himself.

Later, as a young teen, I met another mentor. Gaetano Lombardi owned a small market in Park Slope. After school and on Saturdays, I stocked shelves for him, took deliveries, and helped customers. I'll never forget the day in 1950 he gave me the key to the till. I was sixteen. "I trust you, Charlie," he told me, and it meant the world to me. Again, the most meaningful part for me was that I'd earned his trust. I appreciated the responsibility, and I never let him down. When a man gives you the key to his livelihood, you don't mess it up. Both those guys, Mr. Cavagna and Mr. Lombardi, modeled the opposite of the Italian-American (especially in Brooklyn) stereotype—you know the one. They both worked hard and succeeded through honesty, fairness, and treating people right. I've always tried to rise up to the level of their humble light.

Later on, when I got into engineering and ultimately management, the lessons from my parents and mentors stuck with me. "He's tough," nearly everyone who worked with or for me has said, "but you can trust him." In my book, there's no greater compliment.

So I ask you, what's the good, rational reason a ten-year-old needs his mother to make his breakfast for him? Or fold his laundry? Or essentially solve every dilemma the kid encounters in life, from a bad grade to a bully to a bump on the knee. Maybe we think this expression of our love and empathy and desire to help, which is "natural" and "right," is

actually doing good. But it's not. What we're doing is training a generation to rely on us—the people in charge, the world, the "Man,"—to solve
their problems. Just lie in bed and play your video game, My Lord and
Lady, and eventually, some authority figure will step in and make all
your problems disappear. Whew. And just in time for the bonus round
of your game.

More and more kids these days are diagnosed as learning-disabled.
Regardless of the reasons why there has been such a spike in these diag-
noses, every child deserves the freedom to test his or her boundaries,
relatively speaking. Taking small risks, making their own choices, and
apprenticing at independence. While every situation is certainly unique,
the ultimate goal should always be self-determination. Some adults are
inclined to support a disabled child to a greater extent than the kid actually requires. "We know that sometimes when a child is consistently
overprotected or prevented from taking even small risks, he may learn
to feel helpless or dependent, rather than self-reliant."[6]

The real problem with this approach is that it doesn't foster *trust*.
How can you trust yourself, your commitments, and your promises if
you never have to make any?

In the 1940s, construction all but ceased to provide men and
materials for the war effort. My father, who was too old to serve, was
superintendent of twenty Brooklyn buildings during World War II
(his official title was actually janitor, but today we call them supers).
He depended on my brother and me to help him keep the buildings in
order. In the heart of winter, on Gulag-cold nights, we'd keep the coal
fires burning for the hundreds of Brooklynites he served. He trusted
us to do that job, as hundreds of people trusted him. How common is
that these days, for a kid to have that kind of responsibility—and be that
trustworthy?

Maybe on farms. And, of course, there are all kinds of city, town,
suburban, and rural kids with *too* much responsibility owing to family,
health, or financial challenges. But in the average middle-class family
today, parents practically expect—they trust—their kids *will* crash and
burn—so they might as well just take the wheel.

Our foundation must include trust and self-determination. There is an old adage that we have to pull ourselves up by the bootstraps. Before anyone tosses this book across the room, let me say that of course I know all the challenges in doing so. I understand what the Reverend Doctor Martin Luther King Jr. meant when he said, "That's all well and good, but what about the bootless man?"

Yes, we have gross inequality in our country—in education, housing, income, opportunity. But I'm talking about our *collective* bootstraps. It's up to all of us to do our part with an honest day's work, an honest life's work. Maybe I'm being a little romantic, but you know I've always thought that Mr. Cavagna and Mr. Lombardi came to the ends of their lives of labor and service and wanted for nothing, regretted nothing. Of course, they had terrible days. But they kept their senses homed in the whole time on the far distant reward—leaving their families, their customers, their neighborhoods, their professions, their names, and their country better off than when they arrived.

How is that not a worthy life, well-lived?

Rush Hour

Which brings us to the *long-term* aspect of our private and state remediation project. I call this the 1,776-foot perspective. A national evolution and a major personal change alike require a good stretch of time. You need distance vision to see into a potential future toward which you can and must shift your bones. But we have a tendency in this country—especially over the past fifty years or so—to expect quick-fix solutions to our problems. Paradoxically, we get frustrated when those quick fixes fall apart. For some illogical reason, we anticipate that our temporary solutions will be permanent, or at least semipermanent.

But there are three principal drawbacks to such unrealistic expectations:

1. They don't take the long view—say, the 1,776-foot perspective you get from the top of 1 WTC.

2. They don't often get to the root of the problem—if you leave that underground fire burning, it will eventually consume whatever you put on top.

3. They reinforce and expand our already impatient mindsets—if you ensure that the thing you're building is worth waiting for, then no one will remember how long it took when it's finally done.

This hurry-up/slap-it-up/give-up mentality pervades our personal lives, as well as our national politics and statecraft. We plug holes in the dyke until we're out of thumbs and more holes pop up. After the deluge, we float around from idea to idea, challenge to challenge, task to task, carried along by waves, just trying to keep our heads above water. We're always wondering how we should react to given situations. And therein lies a key mistake: individually and collectively, we're more *re*active than *pro*active. That's because we're doomed to fail in any undertaking that begins with an action, any action, even a "proaction." It's not about the *how*. It's the *why* that matters so much more.

That's counterintuitive, right? Haven't I been recommending that you start this project? Surely a journey of a thousand miles begins with a single step? That's true. But it's not the whole truth. Do we know where we're going, in which direction we're taking that first step and the one thereafter? We might go a thousand miles to some unpleasant, dangerous place. So before the step, we need the road map. That road map is our values. Without preconsidering what really matters to us—identifying and understanding what we want to move toward—we're going to get lost or wind up somewhere we don't want to be.

We all want to see ourselves as people of action—not passive participants in our lives. But there's a difference between foolhardy, reckless action and directed deeds, well-considered. Before we leap, we have to look. Before we dive, we need to assess the depth, the currents, the shark situation. We need to return to a state of what I'd call "eager forbearance." Everything worth doing is worth doing well. We can steadily chip away at what seems to be our most intractable problems only if we aim toward long-term solutions, even ultralong-term solutions we might never experience in our own lifetimes—not just the next step but the ultimate destination.

What's the hurry? I think because of the introduction and steady promulgation of technology in my lifetime, we've become a very impatient people. The idea that I can get frustrated when it takes too long—how long? Twelve seconds?—to download a *Wall Street Journal* article on my iPhone when the Wi-Fi is spotty on the bullet train—this would have seemed preposterous to my younger self.

We have a culture absolutely addicted to instant satisfaction. We have no patience for delayed gratification. We go to the doctor demanding the latest drug for problems now mostly caused by our impatience and excess. We "don't have the time" to take care of ourselves, to eat right and move enough—it's much easier and more efficient from an energy standpoint to max-pack the calories and keep them close by, never getting up from our desk—or worse, our screen.

We certainly "don't have the time" to cure ourselves—we just want to mask whatever problems occur to us owing to this twisted thinking. We're depressed because we feel useless? Tablet. Our blood sugar is high because we eat fast food every day and rideshare up the block? Shot. Our potency is "dysfunctional?" Pill. We spend no time or energy working on the cause of many of our health crises: our lifestyles.

And we do the same with our country. We don't take the time to nurture it, then we grow impatient with it, and finally we give up. We don't plan for the journey. In fact, we spend a lot of the time looking in the rearview mirror and calling the other drivers assholes.

From one perspective, we've become more and more like the eight-hundred-pound person on the scooter, trolling the junk-food aisle at Walmart at three in the morning. We wonder how it is that we got here. In the meantime, what's the point in trying to right the train when it's already gone this far off the tracks? Might as well just enjoy a package of Oreos or two and wait to die.

It's the same with our relationships. Why bother looking up from the screen and attending to the other person when all you'll ever need is in the little box in your hands—and it doesn't expect flowers twice a year? It takes one, two, or a maximum of three words to start or conclude any conversation, to begin to heal any relationship damaged by inattention or recklessness:

1. Thanks.
2. Sorry.
3. Forgive me.
4. I love you.
5. God bless.

Yet how we blather tediously about all manner of bull—or we zip up, cross our arms, and seethe instead. This is the difference between "urgent" and "important" that time management experts have been talking about for decades. Satisfying our immediate hunger feels urgent—taking care of our long-term health is important. Making money is urgent—public safety and your company's reputation are "merely" important. Taking an antibiotic is urgent—keeping yourself healthy in the first place is important. Telling your kids to get their feet off the damn couch is urgent. Telling them you love them? That's important. Firing off that 4 a.m. tweet? We both know where that falls.

Double Your Cookies

It just doesn't seem, though, that we're wired for this kind of long-term strategizing. Various forms of investments in our future elude many of us, and we're worse for it.

In the '60s and '70s, a researcher named Walter Mischel demonstrated this ingeniously in a series of experiments with children. The Stanford Marshmallow Experiment, as it would come to be known, was simple. Kids were given the option of getting a reward—a marshmallow, cookie, or pretzel—right away, or they could delay their gratification for fifteen minutes (during which time the experimenter left them alone with the enticement) and get double thereafter. Know what he found? Those children who were able to delay their gratification—who invested their time, controlled their baser impulses, and focused on the ultimate reward—tended to have better life outcomes. They got better grades. They had better health, got further in school, and so on. They also got twice as many marshmallows.[7] Good things come to those who wait—when they *invest* (their time, money, or attention) in a better outcome. 1 WTC is a towering example of this concept. Yes, we could have slapped together a quick fix on the site. Yes, America had to wait for its marshmallows—but when they came, they were glorious and well worth the wait.

Meanwhile, we're training generations of Americans in the opposite of investing—in expecting, and expecting quick results. Take, for example, the investment of time. When I was a kid in the '40s, if we wanted, say, a Buck Rogers Sonic Ray (with cyclotron chamber and fission speed regulators), we'd have to fill out the form on the back of the comic, send our $2.50, and age out before the toy arrived. This is a bad example. When those things came, they were usually plastic pieces of crap *not* worth the wait. My point is, though, we didn't have such a thing as overnight Amazon Prime delivery (soon coming to you by drone).

Delayed gratification might benefit you, but it ain't easy nowadays to practice it. Kids today can instantly satiate their every whim. I worry about my grandchildren's expectations for what they can have right away with no effort when, while they were still in single digits, they could access all the planet's intellectual resources with a deft shifting of their thumbs. You want to hear a certain song, you've literally got it in your pocket. You need some obscure answer for your homework? The entire compendium of all human knowledge is one Google search away. You need to discuss something with a friend or colleague? Why bother, when you could just send a one-line—one-way—text full of inane acronyms?

Don't get me wrong. I'm no Luddite. I'm excited about how far technology's come and its astounding prospects for the future. I'm particularly enthusiastic about Elon Musk–style open-source innovations that are leading to massive advancements like Hyperloop technology, as well as crowdsourcing and crowdfunding. But there's a disturbing dark side to this latest iteration of in-your-pocket, personal technology. These things act like drugs. They feed our need for the instant fix. They literally work on the brain in that manner. In the same way a bite of a snack cake or a snort of cocaine acts on addiction receptors in your brain that drive you to consume more, your cell phone's ding is designed to get you and keep you *hooked*. That message, that text, that voicemail, that email, that post, that "like" seems urgent. It isn't. And it's almost never important. It just seems important at the time. But it's usually a distraction that interrupts the fulfillment of your core mission—a detour on the way to your destination. With enough of those diversions, you'll be wandering the desert.

There's nothing really new about this phenomenon. Bill O'Reilly recounts that no less than Lincoln himself became "so addicted to the telegraph's instant news from the front that he [couldn't] let go of the need for just one more bit of information…"[8]

Another thing. When I was a boy, we had to engineer and build go-karts from the wheels up for races down our Brooklyn streets. To turn an apple crate, pipes, and string into a "push-o" we had to employ imagination, vision, determination, and endless trial and error to get it right. We learned better that way. If we couldn't stop the thing properly, slammed into a Studebaker, and broke a tooth, that's what compelled us not to give up but to innovate a better braking system. We couldn't just search for step-by-step instructions on a handheld device. Instead, we had to invest enormous time and energy to achieve anything. But that made the achievement all the sweeter.

Sacrificing the definitely important for the ostensibly urgent is always a losing proposition. Investing is *never* urgent. See a pattern here?

Countrywide Adhesive Bandage

This same mentality pervades much of our government. On the national scale, "balancing" budgets and keeping the unwieldy U.S. government afloat is urgent, while re-engineering America from the ground up is important. For the most part, we deal with problems once they're urgent. The I-35 bridge collapsed over the Mississippi River? By all means, let's build another. A thousand bridges will collapse eventually? We'll get to that when we get to it, probably after a bunch more people plunge into icy water.

It's the same in national politics. Our public servants don't often work for their constituents because they share the same values. Instead, if they seem to be getting along at all, it's because the politician is pandering for self-preservation. The pandering targets the public, no matter how ridiculous its wishes; the party, no matter how bullheaded; or the president, no matter how wrong. The instant gratification here is the votes—the next election. We all know that many politicians spend the vast majority of their time running for re-election, even right after they're elected.

There are few principled politicians out there—people like the late John McCain, who always vote their conscience even when it's unpopular, whose values are unassailable and, for the most part, unbending. They take the long view. They're invested in the long-term future of the country, even while keeping a keen eye on maintaining their position of continued influence. These people, despite their fierce ethics and clearly articulated values, paradoxically tend to be the best compromisers. Again, they're focused on the long-term benefits for the whole—not the short-term win for the one.

The quick-fix mentality is certainly the basis for most budget decisions, too. "Balancing the budget," for the few times that happens in a state or our nation, doesn't really mean "balancing." On the day I write this, our national debt is a staggering, seemingly insurmountable $20 trillion—more than $60,000 per citizen—and the federal budget deficit is $593 billion.[9] Our national trade deficit tops $748 billion. But we keep pasting together budgets that deal only with short-term urgencies—not long-term necessities. We keep "kicking the can" down the street to the next generations, borrowing from Peter to urgently pay Paul. We're not taking the 1,776-foot long view that created our nation in the first place.

I'm talking about the difference between band-aids and investments. The difference between slapdash pothole repair and building an interstate highway system meant to last fifty years. As physical health is the underpinning of our personal well-being, federal and state infrastructure is the undergirding of the smooth and safe functioning of our society. In fact, it's essential to remaining economically viable. Infrastructure investment is based in our national values of access, expansion, free trade, and public roads and utilities. When did we lose sight of that? The Republicans (my party), for all their bluster about infrastructure investment, notoriously keep divesting, withholding, and pillaging infrastructure budgets to patch other short-term problems and fund pet projects to pander to their constituents and special interests.

As an example of investment for a long-term mission, consider Social Security. Signed by President Roosevelt in 1935—I was one year old at the time—the Social Security Act of 1935 was substantively revamped in 1983 under my boss, President Reagan. The idea was to create a social insurance program for the welfare of the populace—especially the poor—and provide income to retirees who had worked all their lives. Despite all the liberal yammering about the government's "raiding" that fund, the reality is that the Social Security Administration collected far more in Federal Insurance Contributions Act (FICA) taxes from American workers than it was obliged to pay retirees, especially in the period from the early '80s to around 2010.[10] That surplus contributed to a trust fund of $2.8 trillion,[11] a fund intended to benefit Baby Boomers after they retire.

Yes, most of those funds are invested in special U.S. government bonds, but legally speaking, those bonds will pay the market rate of interest, plus the full principal once they mature. Those funds have not been raided—they're invested. Yes, we must somehow lower overall spending, work on reducing the deficit, and allocate taxes differently or that money will be depleted. But the eighty-four-year-old program, originally envisioned as a long-term solution to Depression-era poverty and insecurity, worked.

You know I don't favor government intervention into our lives any more than is absolutely necessary, so I'm not suggesting some Socialist "utopia" of cradle-to-the-grave services and nanny-state protections. That's not the American way. Of course it would be better for the government to let us manage our own financial savings.

But I concede that as an ancillary benefit of a system in need of dire reform, Social Security has provided real benefits. Ask yourself whether on our own, most of us would withhold and consistently invest in secure bonds as much of our income over multiple decades of our work life as the government did through Social Security to pay for our retirement.

Again, today's expenses are urgent—long-term investment is important. We can argue that the government bears at least some responsibility for the long-term health of our nation, even if it's only for purely self-serving economic reasons.

Now contrast the true investment you've made weekly through FICA since you started working in your teens to the blatant consumerism that, although it's a major driver of a robust economy, benefits your long-term prospects not a whit. The seventy-five-inch curved LCD HD TV you spend $2,000 on when you're fifty is not a necessity—it's a luxury. I'm not saying that we don't all deserve the occasional indulgence within our relative means as a reward for our hard work. But eating when we're eighty—that's a life-or-death necessity. Eating is one of those rare activities that falls in the urgent-*and*-important quadrant. Thanks to the miracle of compound interest, that same $2,000 invested when you were fifty would double its value by the time you're eighty if you just plop it into an account. How much is a thirty-year-old TV—even if by miracle it's still working—going to be worth? How I wish they would teach this to junior high and high school students. How much more valuable is a class in financial investment than some of the more frivolous liberal arts courses now being offered? Seriously, what's more valuable: medieval Portuguese semiotics, or financial literacy?

Groundwork

In any building project, the most important part of the job is the foundation. If the underlying structure isn't rock solid, everything you build on top of, over, and based on it is bound to collapse eventually. It might not happen straightaway—it could occur after many years of normal wear and tear, weathering, or unexpected strains—but a compromised foundation ultimately puts everyone at risk.

We saw this in the collapse of the original WTC towers. The National Institute of Standards and Technology (NIST) concluded that fires intensified by jet fuel undermined the steel structures until the trusses—the long, bridge-like floor sections, inwardly bowed—started to soften then sag. The flagging of trusses transformed their downward pull into an *inward* pull, toward the middle of each tower. This escalating inward tug on the walls eventually caused the outer columns of Tower 2, and later the inner columns of Tower 1, to buckle and fold, and then collapse.[12]

I say take the 1,776-foot long view and hone your vision. But in the late 1960s and early '70s, when engineers, architects, construction workers, and regulatory agencies oversaw the building of the original towers, no one could have reasonably anticipated a scenario in which terrorists would deliberately fly fully fueled passenger jets into the middle of each tower. Right?

Actually, during the design stage of the WTC towers, the Port Authority did, in fact, analyze the impact a single Boeing 707 aircraft might have.[13] We concluded it would likely be local, and not collapse a tower, an event similar to the B-25 Mitchell bomber's smashing into the seventy-ninth floor of the Empire State Building in 1945 on its way to Newark Airport in heavy fog, killing fourteen.

But as the NIST later concluded, we at the Port Authority and our contractors didn't have "the capability to conduct rigorous simulations of the aircraft impact, the growth and spread of the ensuing fires, and the effects of fires on the structure."[14] We didn't even consider any kind of fuel load in our studies. All of that developed much more recently— some of it, especially new approaches to structural modeling, came out of the NIST WTC investigation. We learned many hard lessons that have made more recently-built structures, including 1 WTC, that much stronger. A direct hit from an aircraft 20 percent more massive than a 707, given the momentum, was bound to sever the comparatively flimsy steel of the exterior columns on the impact floors. No one thought to somehow reinforce trusses against flows of molten aircraft aluminum and superheated liquid office furniture and carpets.

But that's not why 2,752 people died needlessly that day. They were murdered by extremist Muslim monsters—not shoddy Western engineering.

Even the best engineering team doesn't have a crystal ball in its toolbelt. We can't predict every single outlandish contingency. Shoddy would be not learning from the terrible lessons. One WTC is one of the safest, most advanced structures ever built. That's thanks, in part, to those who made the ultimate sacrifice. We were able to learn from their terrible deaths.

Now the Port Authority is finishing up the final phase of building the Vehicular Security Center (VSC), the secure compound for truck delivery and underground parking at the new WTC. It's a complex network of tunnels, chambers, and ramps, like a complete city underground. We learned hard lessons after the 1993 attempt to blow up the Towers from underneath. The new VSC's slurry walls are reinforced with seven hundred tons of steel, the weight of about eighteen separate eighteen-wheel tractor-trailer trucks. Each steel reinforcing bar used in the slurry walls is equivalent to one-tenth of the total weight of the Eiffel Tower.[15] The design calls for more than 10,500 tons of structural steel, which is equivalent in weight to about 5,350 average-size cars.

But it's still just a structure based on an infrastructure, built on a foundation. It has vulnerabilities. When the man allegedly said, "Ma'am, even God couldn't sink this ship," it was that kind of hubris that caused the *Titanic* to sink (that and not properly looking out for icebergs, not building the bulkheads up to the ceiling, and a few other engineering blunders).

Does the public marvel at the superiority of the bulkheads in a cruise liner? Nope. But how vital are they? If you were one of the families waiting at what's now South Street Seaport in Lower Manhattan on April 15, 1912, you might say, "Pretty damn important."

I've had the privilege of working on some truly stupendous projects, and I've seen firsthand how the impossible became possible through meticulous planning and careful execution far beneath the gleaming surface. Sure, everyone likes to marvel at the magnificence of a shining tower, a marble arch, or a grand staircase descending from a cruise ship deck. Nothing beats the detailed handiwork on an artful veneer. In fact, when I was a boy delivering the *Brooklyn Eagle* on my Schwinn, the Soldiers' and Sailors' Arch in Grand Army Plaza in the heart of Brooklyn occasioned my first love affair with architecture and engineering. But as I grew up, I took much more interest in what you can't see from a tourist perch. I'll take an ingenious, deep sewer system over a gorgeous façade any day.

The Deep

What the hell does a sewer have to do with America's future? The answer is, plenty. The sewers are a genuinely great place to begin re-engineering, to reconsider from the ground up what we're doing for our country and how we want it to work. Sewers keep our drinking water safe, our streets clean, and our houses, apartment buildings, offices, and factories from becoming cesspools. You rarely see sewers make the front-page news (and when you do, shut your windows and turn off your taps). But you ought to consider them sometime before disaster strikes. Sewers might be invisible to the average citizen, the "butt" of jokes, or beneath the average American's concern. Spoken like a citizen blessed to be in a country where sewer services can be taken for granted. But when a city has to cope with a breakdown in its wastewater system, you'd be amazed at how quickly people tend to take notice.

Superstorm Sandy comes to mind.

If you lived on the East Coast in October 2012, you'll never forget Sandy, one of the most devastating storms to swamp the Northeast in recorded history. It killed 159 people, cut power to millions for weeks, and caused $70 billion in damage across eight states. Sandy also brought into stark relief the vulnerability of critical infrastructure: it halted subway, rail, road, and air traffic. It deluged hospitals, ruined electrical substations, and stopped the water taps for tens of millions of people.

But one of the biggest infrastructure failures occurred belowground: eleven billion gallons of sewage overflowed, a third of it raw.[16] That's some deep…waste.

I've been there. My first sewer job as engineer in charge was overseeing for the city a deep sewer project run by the Tomasetti brothers in Brooklyn. I was twenty-three, and my title was civil engineer assistant, Highways and Sewers Department, city of New York. Working for the Brooklyn borough president's office, I had to approve all the excavations twenty feet under Flatlands Avenue, the sheet piling, the pouring of the pipes. I was on that job for two years.

Later, I became a vice president, chief engineer, and partner for a general contractor firm called J.D. Posillico Inc., a job I held from 1963 until 1981, when I took the job working for Reagan. I oversaw an immense amount of work on sewer lines in those years. It wasn't the only thing we did—in fact, when I started there on November 22, 1963, we were largely contracted to build infrastructure for schools and roads. But we quickly realized the potential of doing the foundational work that nobody sees but everyone depends upon.

There's a moral in that for our re-engineering project on both the individual and larger levels. Find your niche. Focus on the basics. The world feels less insane when you know who you are and what you want to contribute. I didn't do it for the glitz. I just knew it had to be done. It was good enough for my grandfather, my father, and my uncle. I knew my place in the world—and it was good.

Getting Our Wheels Aligned

Let's face it—we're out of whack. We feel out of balance, harried, burned out, and confused more than ever before. In short, we're overwhelmed. We find ourselves putting out fires rather than spending our limited resources on fire prevention.

There are two problems here. The first is that instant fix we seek, about which I've already spoken. The second is more complex. Our behaviors—our thoughts, decisions, and actions—are often not aligned with our *values*. That's if we've even considered our values at all. I propose that we absolutely must get back in touch with our values—and ensure they're good and right and true. As forerunners of our own futures, guardians of our families, and stewards of our country, we must utilize and continually hone a values-based leadership paradigm.

What does that mean? It means before we set out, we align our steps with our programmed destination. Then we go forth boldly, confident in what we believe, what matters most to us, where exactly we're trying to get. We allow those values to drive our behaviors and choices toward a deliberate and decent direction. In this way, we can steer our own ship. No matter which way the wind blows, we control the rudder. The world around us—circumstances, theories, challenges, politics, economics, and so forth—might change, but we stay true to our course. We're following a compass point we predetermined. The late leadership expert and educator Stephen R. Covey calls this point "true north," and he recommends we don't deviate.

But where is that point? There is no actual compass, right? That's somewhere inside us, somewhere theoretical, invisible, certainly subjective? Well, that's half-true. The first step in values-based leadership is self-reflection and self-awareness: you must search both inside and out, explore, and clarify who you are and what you stand for, what you want to move toward. "After all, if you aren't self-reflective, how can you truly know yourself? If you don't know yourself, how can you lead yourself? If you can't lead yourself, how can you lead others?"[17] If we can't lead others, we certainly can't expect America to lead the world with light and truth anymore.

Consider the nearly unlimited values on which you can base your personal life, career, and contribution: Power. Money. Humility. Trust. Health. Cooperation. Diligence. Loyalty. Love. Control. So what motivates you? For this self-inventory, you might consider:

- What matters most to you?
- What brings you peace, satisfaction, fulfillment?
- How do you define "success"?

- What's your character?
- How would others describe you?
- What would you want people to say about you in your eulogy or obituary?
- Do you think God or karma or some other power greater than yourself put you here, now, for some specific purpose, which some people think of as a "calling"?
- What character traits do the people you most admire from history and your own circle share?

I like Covey's idea of true north, but I'd prefer a term with more resonance for us as Americans. Despite the controversy of the term—lots of people don't care for it—I prefer "Ground Zero Values." Here's what I mean. Think about the way you felt on 9/11 and in the hours, days, and weeks after the attacks. Didn't you want to hold your family closer? Didn't you want to put up a flag? Didn't you feel less of an Iowan or Mainer, less of a Trentonian or Denverite, and more of an American? Didn't you feel more connected to God and your neighbors?

Now think about the first responders, the workers, all the volunteers who rushed to the site and stayed until some of them literally couldn't breathe anymore. What did they value? Was it the fanciness of their cars? Their kids' soccer rivalries? Their latest snarky social media post? No. They cared about coming together, serving their fellow humans, honoring the dead, helping families find some measure of peace, putting their backs into hard work that had to be done. Times like that test people's resolve. You see what they're made of, their values in action. It's the Ground Zero of what they really believe and care about.

I was in the room on September 13, 2001, when Pataki and Giuliani spoke with the president again. I will never forget his words:

Well, thanks, Rudy, and thanks, George. Let me make it clear to you all as my close friends that my mindset is this: one, I weep and mourn with America. I'm going to a hospital right after this to comfort families. I wish I could comfort every single family whose lives have been affected. But make no mistake about it: my resolve is steady and strong about winning this war that has been declared on America. It's a new kind of war.... And this government will adjust, and this government will call others to join us, to make sure this act, these acts, the people who conducted these acts and those who harbor them, are held accountable for their actions; make no mistake. And as we do so, I urge—I know I don't need to tell you all this—but our nation must be mindful that there are thousands of Arab Americans who live in New York City, who love their flag just as much as the three of us do, and we must be mindful that as we seek to win the war that we treat Arab Americans and Muslims with the respect they deserve. I know that is your attitude as well—certainly the attitude of this government: that we should not hold one who is a Muslim responsible for an act of terror. We will hold those who are responsible.[18]

Wow. It's hard for politicians to impress a cynical old-timer like me—but W really blew me away that day. His words made going back to the devastation just a little more tolerable. They summed up the Ground Zero of our national values. That's leadership.

Like that of my friend, Doctor Michael Smatt, who immediately met with FEMA after 9/11 and secured credentials for chiropractic professionals to have full access to Ground Zero. With the New York Chiropractic Council, FEMA, and the Red Cross, he coordinated several sites at Ground Zero to provide chiropractic care to police, firefighters, and rescue workers twenty-four hours per day, seven days a week, for one full year.

My daughter, Carla, a massage therapist, was there too, donating time to the guys at Ladder 5 and other workers. Thousands of citizens did what they could.

Value Added

Identifying and clarifying your values—and I recommend writing them down in a personal mission statement—will put you among the top 1 percent of the top 1 percent of leaders. Almost no one does it. But everyone who was ever anyone most definitely has. For examples of values-based leaders, think Churchill. Think Lincoln. Reagan. Susan

B. Anthony. Golda Meir. Jesus. Joan of Arc. Mary Magdalene. Harriet Tubman. Rosa Parks.

On the other side of the coin, Hitler. He knew his values. He wrote them down. He managed to seduce millions into unspeakable acts through his sheer confidence in those values and ability to communicate them, inspiring others to march lockstep toward a fiery terminus. How many politicians have values that lead their decisions and votes? I'd rather have a leader with values I don't at all agree with—not Hitler, of course—than one who has none at all.

If you're having trouble defining your values, try this (though it can reveal unpleasant truths). Reverse-engineer it. Over the next few days, monitor your behaviors, how you allocate your time, what you think about most often, what unfulfilled aspirations nag you. What you're doing right now with your time, money, and mind perfectly reflects what you value. If you don't like what you uncover, reboot and redirect your behaviors to reflect your real values. Otherwise, accept your values for what they are and stop feeling guilty. Keep doing what you're doing and be proud of it. Ideally, probably some combination of both these techniques will work. This is an opportunity to alter some of your behaviors that are incommensurate with what you value, and also to discover the real reason for other behaviors that have long bothered you.

Another technique I've used and know other leaders use is to consider the question, "What would I do with my time and other resources if I had just six months to live? Six weeks? Six days? Six hours?" Wasn't that going through all our minds on September 11? It's illustrative, if agonizing, to listen to voicemail and read text messages of people on the planes and trapped in those fires that day, knowing they might have six minutes or less to live. Knowing that these might be the last words they ever said to their loved ones. Their words are what they considered in their last moments to be the most valuable thing they could say. Their words are comforting. Reassuring. Cautioning. Loving. Candid.

"The only thoughts I have are of Nicholas, Ian and you. I am terrified. I needed to tell you that I truly love you. Always, Diane."

"Honey, wanted to tell you how much I love you. I was a little worried. I don't want to lose you now that I got you back. You mean everything to me. You have my whole heart and life. I love you so much."

"I want to hold you now."

Those messages[19] are heartbreaking but also heartwarming. They're perfect examples of Ground Zero Values. Employing this thought experiment in an honest and attentive way will tell you a heck of a lot about yourself. It can also be a stark reminder of how changeable our values might be. If, God forbid, you found out today your child was dying, wouldn't that bump your work project or the broken sprinkler down your list of priorities? It ought to. So why would you program your internal GPS away from your children without that happening and toward something else? What are you waiting for?

Probably the single most important thing you can do for your time management, life balance, personal fulfilment, and national services is to ask yourself the following question many times a day, answer honestly, then act accordingly:

"Is what I'm doing *right now* the most valuable use of my time?"

Time and motion expert Brian Tracy, who has advised politicians and business leaders for decades, summarized a generation of research into that one question. Remember that "valuable" is defined as "according to your values," in accordance with your mission, what you're trying to move toward and get more of. Always concentrate on those things that are *important*—but *not urgent*.

As Americans, we should learn to delay our gratification in order to get the two marshmallows instead of one. We should invest our time, energy, money, and minds into our personal and collective futures. We should take the 1,776-foot long view, even the ultralong view that goes beyond our own lifetimes.

Posteriorities[20]

When your beliefs, values, and behaviors are all in line, it's called balance. You're in sync. You feel a kind of peace you're unaccustomed to. You never struggle with being overwhelmed because you've already decided in advance what things are more important than other things—where you're going in life and which way.

To use a simplified example, let's say that as a student, you value the STEM subjects (science, technology, engineering, and mathematics) over the humanities. Specifically, you strongly prefer math over English. I know I was like that back in school. If I had homework in both subjects, I let myself be okay with spending more time on the math. It was more interesting to me, more satisfying, more fun, more *valuable* than English in terms of my mission, what I wanted to do with my life. I guess you could say I valued numbers more than I valued words.

Or let's say, later in life, you clarify that you value being there for family above satisfying your boss. Is that all right? Of course. They're your values. The good news is, you'll never feel torn between the two when they're in conflict. In fact, there will be no conflict. You won't feel pulled in both directions. Work obligations don't actually compete with family commitments in your case. You're not trying to decide whether to stay late at work or go to your kid's ballet recital. You're always—within reason—going to be there for your kids.

That means you know in advance that sometimes you won't be there for your job, career, boss, or customers. This kind of self-confidence about what you value also requires you to accept the consequences of acting on those values. Heading toward a particular place on a map obviously means you're heading away from other places and skirting still more. In this specific case, maybe it takes you longer to get the promotion. Maybe you make a little less money. But I've found that rewards—financial and otherwise—tend to favor those whose values are more stabilized in Ground Zero than propped up on ramshackle shaky ground. If you have a boss or work for an organization that would penalize you for conflicting values related to your family, your health, or your faith, consider whether you're in the right place.

In the private sector, I've had as many as eight hundred people working under me, and in public life, ten times that. But I don't recall ever having insisted that someone prioritize me over their spouse, parents, or children. And I've worked for some of the modern era's most powerful leaders, yet I don't recall them ever demanding I choose them over my family, my integrity, my health, my religion.

Back to the school example. When I decided to devote more of my personal resources to math and science, I had to be okay with maybe getting a lower grade in English. If I considered both those subjects equally important, I might have become overwhelmed instead of having a reasonably good experience in school. You can't get an A in everything. We're living in a time and place where you get to choose the subjects you want an A in, and those you're willing to get a C in.

I like to golf. But I don't have to be a better golfer. I don't care about how well I can golf and how I might measure up to others—even the president. But I do want to be the best grandfather I can be. That matters to me. I value leadership, but I value contribution more. That's a major reason I'm writing this book. I value both of those things over, say, my ability to dance (which I used to care about very much, mainly because I cared very much for my first crush, Prudence, and *she* cared about dancing).

Now, of course, I also value excellence overall, so I don't want to do anything I stink at, and I try to design my life that way. I'm not saying I ignore those things I don't value, but I don't seek them out, and I certainly don't tear my hair out trying to master them. But overall excellence, execution, stick-to-it-iveness—that's the only way we were able to devote our entire lives for more than a decade to rebuilding the WTC.

Finally, this process is necessary not only for time management, life balance, and project fulfillment but for morality. I believe God has imbued us with an internal compass and sacrificed his own son so we'd have external directions, too—a pilot, a guide who traveled the right road in the right direction. Maybe you have other models, other maps, your own private North Star. The important thing is that to keep your hands on the ball, you have to ignore your untied shoe. You have to focus on the important rather than the urgent.

Ask yourself, "Is what I'm doing now the most valuable use of my time?" If not, adjust your life accordingly. Daily. Hourly. How much time did you spend today surfing the internet or trawling through your emails? How much time did you spend on your life purpose or helping others?

Focus like a laser and concentrate single-mindedly on those behaviors and actions likely to earn you the highest return—those things that match your values. This kind of values-based life assessment is one reason it took so damn long to get the new WTC built. We valued getting it right. We valued public participation. We valued collaboration across fields and industries. We valued partnerships. If we'd caved to the superficially "urgent" demands of impatient citizens, critical politicians, and a goading press, we would not have built the safest and most elegant and graceful structure in the world—and the most *important*.

Mission Accomplished

How do we expand those values that rebuilt Ground Zero to the more formidable and important project of re-engineering America? We start with the shared belief that service should be at or near the top of all Americans' lists. With notable exceptions, no one participated in the rebuilding of Ground Zero for the fame or the money or the fun. It was a grueling, nail-biting, problematic *ordeal* from start to finish. We all made the project tolerable—totally doable—not by avoiding it but by digging in. To paraphrase President John F. Kennedy, we didn't do it because it was easy; we did it because it was hard. We all, from the humblest scaffold mechanic to Governor George E. Pataki and President George W. Bush, shared one common ethic—a belief that despite the odds, we Americans could get the job done, could make the impossible viable.

Let that effort stand as a model for what to do when circumstances—financial, familial, terroristic—force you to take part in something unpleasant. Dig deep and search your spirit for some value you have that the activity touches upon—then ride it till the end. Maybe it's duty. Maybe it's standing up to a bully. Maybe it's nothing more than hope.

I'm afraid a lot of Americans have given up hope. For various reasons, some legit and some ridiculous, many of our fellow citizens no longer believe in American excellence. I still believe. I value the *idea* of America—and our *actual* America. All those who have ever gone to war or into public service, or who have sacrificed important things like family and fun for the greater good, embraced that concept. They value their country in some cases more than their own comfort, more than their individual success. That's the only way to win a war, by the way. Of course, you'd rather be watching your kids race go-karts—but there's something larger at stake. An intangible principle. This greater ideal dovetails into the other things you value. Your kids won't have their freedom, safety, or comfort anymore if you don't spend this time away from them, doing this important thing. Think about those few things that would motivate your willingness to go to war.

There was no one I knew on the WTC project who didn't share the value of resiliency. Of true grit. Of American exceptionalism. Since 9/11, as a nation, we've collectively valued protection against terrorism more than we've valued individual privacy. The passage and continued success of George W. Bush's Patriot Act—one of the most valuable tools in our arsenal against our enemies—has proved that.

We value thoroughness over convenience at airports. But there's a limit, don't you think? Would you be willing to have soldiers come to your house, seize your property, rifle through your stuff without a warrant, detain you on a hunch—even if that did mean we as a people might be safer? Of course not. Luckily our Constitution already prevents that from happening. Those values are imprinted on our founding documents, interwoven into the fabric of our laws—they're inextricable from who we are and why we are.

This is not to say it's always simple. What to do about issues like the Syrian refugees is a complex, multifaceted problem. We value liberty, don't we? Bring 'em in! Bring me your tired, your poor, your fleeing-from-monsters. But we also value self-protection and national security. What if one of them wants to take advantage of our hospitality like Aesop's scorpion, who killed the frog after the frog carried the scorpion on its back across the river—just because it's the scorpion's nature? It has certainly happened before. We need to decide together what we believe and how to act accordingly, based on our collective values.

Again, we can look at the Ground Zero rebuild as an exemplar of how to accomplish a mission with important and seemingly conflicting values such as security and commerce, whose purpose is both to memorialize our fallen and to move us forward.

3

Model and Localize

Secure Your Own Mask

How you treat others is the true measure of who you are inside. This is applicable for us as individuals and as a nation, too. Only after you've secured your own oxygen mask are you ready to help others with theirs. Only after you've gotten your own affairs in order are you able to become of real use to those with whom you interact at home, at work, and in service beyond. You'll be of no use to anyone if you choke from lack of oxygen.

For the same reason, you need to take care of yourself as a parent. Family ought to be on every American's "important" list. Putting your kids first, though, sounds good on paper and in a bar conversation, but those who take that expression literally every time are doomed to orphan their children. The most important thing you can do for your children isn't feeding them only the choicest cold cuts and organic peanut butter, or sending them to the best ski camp in the costliest gear. The most vital thing you can do to secure their future happiness and stability is to model for them the kind of people you wish them to become.

My mother, Grace (Morringiello) Gargano, may she rest in peace, was such a parent. She was a model and leader extraordinaire, the best provider of love and support I've ever known. I'm not talking about financial support, though that can surely help. It doesn't mean whatever your kids want, they get. I'm talking about affection, protection, connection, and correction when necessary. I was my mother's prince. I never left the house without kissing her on the cheek, never came home from school without kissing her on the cheek. My brother Tom would tease, "Oh, look at sweet little Charlie, he's kissing Mommy." But that's what I had to do. I had that need.

I sometimes allow myself to wonder whether my mother's tight apron strings on me stemmed from her loss of not one but two children before I was born. This is hard to talk about. Gianni, named for my father, died in a freak accident when he was a toddler. He and my brother Frank were playing tug-of-war with a rag outside the house on my grandparents' farm. Somehow one of them let go, and Gianni plunged into a tub of boiling water my grandmother was using for the wash. Then Frank's twin sister, Teresa, named for my grandmother, died in Brooklyn of pneumonia. We didn't have heat in those early days, and the kids did not fare well on those trips back and forth to Italy. Poor Frank.

Meanwhile, Tom was never really a brother to me. Remembering him reminds me that the modern family is an institution in need of re-engineering. I agree with the ostensibly pro-gay slogan of the '80s and '90s that "love makes a family, nothing more and nothing less." It's a very good beginning, love. But I think there are other things you also need. I think you need a father and a mother. Hypothetically, you could have a good deal of the right kind of modeling from same-sex, committed, responsible, stable partners. But it's not the ideal. Gay marriage is not necessarily part of the erosion of the traditional family I'm worried about, but it's *not* a traditional family in the way I know it. I'm not opposed to it. I know most of the modern literature makes it clear that the gender of the parents doesn't have much to do with parenting abilities and outcomes for kids. But I think children miss out when they don't have models of both genders. Then again, I guess there is no perfect way to make a family.

Nourishment, Italian-Style

I spent my first four years with my father's parents in Southern Italy, where I was born during one of my parents' sojourns. Life centered around family—and family centered around food. You can imagine the food. My grandmother always had fresh pasta, fresh tomatoes, farm-fresh cheese, homemade bread, even a little wine for the kids. Look at those carbs! So why is one of the world's five "Blue Zones"[1] (where people live the longest) in Southern Italy?

Among the reasons are a focus on family, moderation, life purpose, and judicious regular exercise.

In short, values differ among the longest-lived and the sickest people. In many cases, people with lifestyle diseases—the "diseases of kings," such as diabetes and heart disease—are sick because they've made certain choices that in our culture are common and hard to avoid. Most of us value instant gratification—a quick "hit" of high-fat, high-sugar, high-cholesterol, and high-salt convenience food. Or we're looking to fill soul-level holes—where our life purpose should be—with pizza and sweets. And then we wonder why we get out of shape, get cancer, and die young.

Instead of changing our behaviors by reflecting on our values, we seek the very thing that got us into trouble in the first place—the fast fix, the soul-hushing in the form of pharmaceuticals. You can bet Big Pharma is in love with this kind of seeking. For all their scientific and medical advances, such as insulin and the polio vaccine, those companies value their bottom line above all. Their investors want their dividend checks this year—damn the future liability when every other drug they promulgate turns out to cause anal seepage and death. If a drug turns out just to chase the tail of the problem, ultimately making it worse, hey, all the better for that bottom line.

Family's the best medicine. Food in moderation is medicine. Working in the fields (or whatever metaphorical fields encompass your chosen "field") is medicine. Purpose. Do you think anyone at the Ground Zero site complained of the sniffles?

Growing up, it might have seemed urgent to find our stickball, lost somewhere in the trees of Park Slope after a fly ball on a Sunday afternoon. But when my sister Connie called out the front window at ten minutes to one in the afternoon—when my father was ready for my mother's cooking—it was important to be seated at the table by 1 p.m. sharp—often with half my friends in tow. See, my mother would make enough for sixty of us on a Sunday. My Irish friends would all ask to come, because they would otherwise be eating meatloaf and catsup at five o'clock. This was the prescription, the daily inoculation against the contaminations of the outside world.

I have a big, beautiful photo of my parents cheek to cheek. I keep one copy of it in my apartment in the city and another in my home in the Hamptons, so I see them every day, every night. My mother and father were magnificent people. They taught us a lot. My mother thought I was the best thing since meatballs, but my father was on the tough side. He thought she was overprotective, smothering me, letting me get away with murder. As a result, I didn't have a good relationship with my father. I didn't meet him until I was four and a half years old because, when I was one, he left my mother and me in Italy and went back to the U.S. to get us started up again. He took with him my older brother Tom and his own younger brother, my uncle Peter. With the war beginning to rumble through Europe in '39, he hustled us out and back to the U.S. in June of that year.

What a journey. We took the Italian Line flagship steamer, the SS *Rex*, out of Naples. At 880 feet and 51,062 tons, the *Rex* was the same size as the *Titanic*—and, boy, were the Italians proud of her. We sailed tourist class among four hundred others—there were more than two thousand passengers all told—for nine days. I was ten when my father, with a rare tear in his eye, told me the British had sunk the *Rex* so the Germans couldn't use it to block the harbor in Trieste. She lay capsized and burning for four days before she finally sank off the coast of Yugoslavia. I saw her picture in the paper and got a lump in my throat—it was surreal. My father nodded his understanding. An era had ended—another had begun.

The Patriarch

Like his father before him, my father, Giovanni "John" Michael Gargano, was a classic patriarch, with all the positives—and some negatives—that entails. He was born in Brooklyn in 1904, while his Italian father, Carmine, worked the season in America. It was the year the New York City subway ran its first underground train. The Williamsburg Bridge had just opened in the shadow of that other famous bridge. Theodore Roosevelt was president. The Dreamland amusement park premiered on Coney Island. Ebbets' team, the Brooklyn Superbas—officially the Brooklyn Base Ball Club, which would evolve into the Dodgers—finished sixth in the National League. And Kings County, synonymous with the borough of Brooklyn, reigned as the horse racing capital of America. The Gravesend Race Track was the mother of all venues, comprising the vast tract between McDonald Avenue to Ocean Parkway, and Avenue U to Kings Highway.

But my paternal grandfather, Carmine Gargano, like most working-class New Yorkers, didn't have much time for amusement during his stints in America. Those days were tough for the nearly four million New Yorkers. Although the city was still about 98 percent white, more than a third of the population was foreign-born—still is today. But back then, there were also still two hundred thousand horses in the city. There was so much manure, the populace happily exchanged it for motor vehicle exhaust.[2] Many residents still used outhouses. There were rudimentary sewers, but many buildings didn't connect to the mainline.

An Irish immigrant cook named Mary Mallon had been unwittingly spreading disease across the city. But "Typhoid Mary" was not unique. To combat outbreaks of typhoid, smallpox, diphtheria, malaria, yellow fever, cholera, tuberculosis, and other deadly scourges, New York City took the lead internationally in sanitation—with street sweeping, modern sewers, and massive water mains. This meant jobs. And that meant immigrants.

Twelve new immigrants arrived at Ellis Island every minute.[3] Each year, my grandfather Carmine was one of them. In the nineteenth century and the first decades of the twentieth century, waves of Santangelesi traveled from Sant'Angelo to South America and through Ellis Island to Brooklyn and some points farther, such as Pennsylvania, to escape poverty and persecution[4]. He made the annual pilgrimage from Naples to New York, a two-week trip at the time, third class. In 1904, he arrived with his pregnant wife, Teresa, aboard a little armed boarding steamer, the SS *Italia*, out of Nantes, France, on one of its first voyages. That was the height of Italian immigration to the U.S. Between 1900 and 1914, 3.5 million Italians crossed the Atlantic.[5] (The *Italia* would be torpedoed and sunk in the Mediterranean by an Austro-Hungarian sub in 1917—someone didn't like all these Garganos coming to America!) On these annual journeys to New York City, Carmine worked the four or five winter months in the city's underground.

In 1907, the city tapped the Catskill region as an additional source of water. This meant building new aqueducts, tunnels, reservoirs, pumping stations, and treatment plants. In the following few years, men like my grandfather built the 163-mile aqueduct, placing it between 174 and an astounding 1,187 feet underground. They built fifty-five miles of cut-and-cover aqueduct, nearly thirty miles of graded tunnel, thirty-five miles of pressure tunnel, six miles of steel siphon, and thirty-nine miles of conduit. Nearly half the city's residents still get their water through my grandfather's *acquedotto*. He was justifiably proud of his early accomplishments, as am I. From 1842 to the present day, there have been no significant interruptions in water service to New Yorkers.[6] That's amazing. Even during the Civil War, New Yorkers could turn on their spigots and get fresh water—many say the best in the world.

When he wasn't working on the Catskill Aqueduct, my grandfather laid Orangeburg pipe—bituminized fiber pipe made from a mishmash of wood pulp and pitch, which served most citywide water and waste transport for more than a century before PVC made its debut in the 1970s.

He was a tough guy, my grandfather, which is where his son John got it from. One of my earliest memories—I must have been three—is of my grandfather shooting our sheepdog under the apple tree on the farm. "Once these bastards get a chicken, they never lose the taste," he said. Eventually, he became a foreman in the water mains in New York. And like many of his compatriots, he had to carry a gun while underground so nobody would mess with "the boss." It was a little bit like the Old West, I've always thought. At once booming and a little lawless.

But in those days, despite the stereotypes, there was surprisingly little violence between immigrant groups. A little anti-Semitism (a little is always too much). Some teenage turf scuffles that ended with bloody noses and battered egos. Some remnants of the infamous street gangs of the nineteenth century—the Five Points, Dead Rabbits, and Whyos. And the inklings of what would become the organized crime syndicates that took off in the mid-twentieth century.

They continue to this day. Typical *Daily News* headlines that make me gulp run almost daily: "Carmine (The Gorilla) Gargano Jr. [*no relation!*] wouldn't stay down even after his Colombo family mob buddy put a bullet in his body and another in his eye."[7]

There's almost never any news about regular, hardworking, Italian Catholics—the vast, vast majority of us. For this reason, I sympathize with the majority of Muslim Americans trying to go about their day, feed their kids, and get them off to school, while their neighbors assume they're putting together bomb vests in their basements.

No, the way my grandfather told it, perhaps through the lens of wistful reminiscence I've now come to understand as the province of the elderly, everybody got along. They understood they had a job to do, and they had to work together to get it done. When you're in the sewer together, color, creed, and national origin tend to get covered up in muck. The same shit sticks to the Poles as to the Italians, to the Jews as to the Irish Catholics. And maybe that pistol helped.

It was the same thing on Seventh Street—minus the pistol. We were mostly English and Irish kids, with a few scattered Italians like the Garganos. What mattered to us was the pride we had in our block, how we were going to be the best stickball players and defend each other's honor against those Ninth Street bozos, those Fifth Avenue goons.

Familia Gargano

My father was born into U.S. citizenry. My paternal grandparents took him and his younger brother, my uncle Peter, back and forth between Avellino and Brooklyn. In his mid-twenties, after he returned to America with his new wife from the old country, he decided to follow his father, Carmine, and work in construction.

Bad timing. It was 1929. The Roaring Twenties came crashing down that year onto the corpse of an eerily silent Wall Street. The groundless exuberance in building development that had characterized the city for decades went poof on a Tuesday in October:

> For real estate developers, 1929 seemed like a slot machine stuck in the pay position. In Manhattan in 1928, plans were filed for 14 buildings of 30 stories or higher, but by the next year the number was 52. Only with the passage of time did it become clear that of these 52, just 19 would be built.[8]

One of those ambitious plans was for the Empire State Building, which forever changed—if not defined—the cityscape of New York. Another was the palatial new Waldorf Astoria, the largest and tallest hotel in the world at the time, and still the brick embodiment of New York City affluence. The New York Central Building, now the Helmsley Building, at the foot of Park between East 45th and 46th Streets, is a beloved landmark now, especially at holiday time.

But other grand designs—including a one-hundred-story skyscraper for the Metropolitan Life Insurance Company—foundered for money or toppled after financial backers lost their shirts and nerve.

Underground, the water and sewer infrastructure projects slogged on. My father worked through the '30s on such projects, following in his father's waterlogged footprints. The Depression that followed the Great Crash led to a renaissance in infrastructure money. FDR funneled a lot of money all over from Washington, and New York state channeled more down to the city. New York had suffered terribly from the loss of billions, but it was a major beneficiary of the New Deal. The Works Progress Administration (WPA) was busy putting tools in hands, money in strong boxes, and chickens in every pot by funding projects like the sewer reconstruction on 123rd Street in Queens and on 125th and 179th Streets in Manhattan; the 65th Street Transverse sewer that crosses Central Park; court buildings downtown; parks and the Arthur Avenue Retail Market in the Bronx; playgrounds, roads, and schools across the city.

My father did well. He rose to foreman by his mid-thirties. He was able to buy a tall brownstone at 414 Seventh Street—the house I grew up in with my whole family, including my uncle Peter. He then bought a bunch more brownstones across Park Slope.

And then the war broke out. Most construction stopped again. All the government money that had been building America went instead to the effort to destroy Germany and Japan. All those men and women tasked with building went overseas to destroy or be destroyed.

Don't get me wrong. Never before had there been a greater cause. It's just a shame that we always look at these two things in such a binary way. You put your money either in defense or in infrastructure. We never seem able to do both.

My father, born and raised on construction and too old for the draft, had to find work, so he became a janitor. He worked for building owners, primarily in Park Slope, taking care of odd jobs but mostly shoveling pulverized coal into boilers so the residents could stay warm and enjoy one of the spoils of modern life—hot water.

When you think of Park Slope, you probably think of brownstones on the streets. But on the avenues—Seventh, Eighth, Ninth, and Prospect Park West—there were a lot of big apartment buildings. And those buildings had to be taken care of, heated, kept running, and cleaned. With fifteen or twenty buildings all in need of coal every night, my father needed to enlist a crew. Guess who? Me. And my older brother Frank.

Remember, I'd grown up in Italy with Nonno Carmine and Nonna Teresa. I hadn't even met my father until I was four. We had an uneasy relationship. We were strangers, after all. After school on Wednesdays, Thursdays, and Fridays, and all day on Saturdays, Frank and I worked my father's crew. After delivering the *Eagle* for Mr. Cavagna and stacking

"North Tower Burns, 10:15 a.m., 9/11/01" – Moments after the South Tower collapses, the North Tower continues to burn and produce mass quantities of thick smoke. Thirteen minutes later the North Tower would collapse.

Images & Text by Gary Marlon Suson

"FDNY Chief Zaderiko" – FDNY Battalion Chief Stephen Zaderiko. Operations Chief at Ground Zero for nine months. Lost sixty-six of his friends on September 11.

"FDNY Capt. John Vigiano" – FDNY Captain John Vigiano, Sr. (1938–2018) cradles the American Flag used for his son's funeral. John lost both his sons on September 11, NYPD Detective Joseph Vigiano and FDNY Fire-fighter John Vigiano II.

"Dennis O'Berg and the Harry Potter Book" – FDNY Lt. Dennis O'Berg, who retired on the morning of 9/11 so he could dig full-time for his missing son, FDNY Firefighter Dennis O'Berg, Jr., cradles the Harry Potter book his son read to his children the night before 9/11. His finger marks the last page that was read.

"FDNY Father Chris Keenan" – FDNY Father Christopher Keenan, chosen as the FDNY Chaplain when Father Mychal Judge died in the collapse of the World Trade Center, stands in the dirt of Ground Zero below the WTC Cross moments after the weekly Sunday prayer service was held for Ground Zero recovery workers.

Images & Text by Gary Marlon Suson

Images & Text by Gary Marlon Suson

"FDNY Chief Turner at WTC" – FDNY Battalion Chief Bobby Turner stares at an area of Ground Zero, winter, 2002. Chief Turner would later go on to become the FDNY First Deputy Commissioner at Metrotech Headquarters.

"FDNY Capt. Joe Downey holds Dad Ray Downey's Dress Cap" – FDNY Capt. Joe Downey stares off in thought toward the WTC site while holding the recovered dress cap of his father, Deputy Chief Ray Downey, who died on 9/11.

"Big Frank and the WTC Cross" – Ground Zero construction worker Frank Silecchia stands below the World Trade Center Cross, which he discovered Sept. 14, 2001 in the rubble, in winter 2001. The "WTC Cross," two giant steel beams that broke off from a high floor and landed upright in the rubble in the shape of a cross, became an iconic artifact of the recovery period.

"The Frozen Clock" – Iconic image of a Timex clock discovered far below Ground Zero in the PATH subway tunnels, frozen at 10:02 a.m., the exact time the first tower collapsed, blowing out all the electricity to the area and marking the deaths of at least 1,500 people.

Images & Text by Gary Marlon Suson

Images & Text by Gary Marlon Suson

"ATV Mike" - Mud-caked 9/11 Recovery Worker Mike Bellone. Spent nine months digging for victims with the FDNY WTC Task Force, despite losing both his parents during the Recovery Period. Bellone is now chronically ill with a 9/11-related heart condition.

"Old Irish Ironworker" – A 9/11 Ironworker from Ireland lifts his mask for a smoke break at 2:00 a.m., winter 2002.

"Oscar Prays at Sunrise" – Gary Marlon Suson's iconic image seen around the world of FDNY Firefighter Oscar Garcia appearing to say a short prayer prior to beginning his dig for missing 9/11 victims at 7:00 a.m. on a spring morning, 2002.

"Wheelchair Salute for Fallen" – A disabled FDNY firefighter salutes the body of a recovered FDNY firefighter as the body is brought up the exit ramp during the "Honor Guard" procession in spring 2002.

Images & Text by Gary Marlon Suson

Images & Text by Gary Marlon Suson

(Left) "FDNY Firefighters Dig for the Missing" – On a cold winter's day at Ground Zero, 2001, firefighters dig tediously for the missing victims of the September 11, 2001 attacks at The WTC.

(Bottom) "The Hole at Ground Zero" – Photographed from where the South Tower once stood, we see the many layers of twisted metal and concrete going down one hundred feet from ground level. The water seen here had leaked in through the cement walls that were erected in 1966 to keep the Hudson River out.

"Ground Zero Seen from the PATH Train Station" – From the unique perspective of the destroyed PATH subway train area, we see Ground Zero on a rainy spring day. March 2002.

"FDNY Prayer for Recovered Firefighter" – With the skyline of New York in the background, FDNY Firefighters and a Chaplain say a private prayer for a recovered FDNY firefighter at Ground Zero, spring 2002.

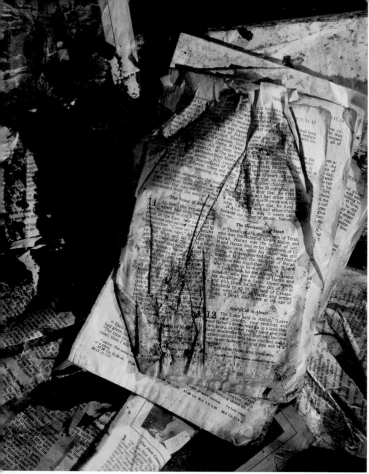

(Left) "The Bible Page" – Official Ground Zero Photographer Gary Marlon Suson's rare discovery of a lone, wet, charred page from the Bible containing the verse Genesis 11, about the Tower of Babylon.

(Bottom) At an event at FDNY Ten House on December 4, 2004, where the strength and courage of all the first responders filled the room.

Image & text by Gary Marlon Suson

Below photo by Lester Millman

(Top) Busy planning the rebuild in February 2002. *Photo by Lester Millman*

(Bottom, from left) The author, NY/NJ Port Authority Executive Joe Seymour, and Gov. Pataki, in a portion of the PATH train station destroyed during the September 11th attacks. *Photo by Lester Millman*

(Top) Gov. Pataki and Chairman Gargano in agreement, as usual, outside of the James Farley Post Office on October 8, 2002.

(Bottom) Chairman Gargano enjoying good company at the James Farley Post Office on October 8, 2002.

Photos by Lester Millman

(Top) Shovels are out and the new construction begins at Battery Park City on August 20, 2003. *Photo by Lester Millman*

(Bottom) Tough decisions to make during a meeting at the temporary Port Authority office in May of 2004. *Photo by Lester Millman*

(Top) I had a good wingman. Chairman Gargano speaking at Daystar Tech Industries on June 21, 2004 as Gov. Pataki looks on. *Photo by Lester Millman*

(Bottom) Making the pledge to recover at the cornerstone ceremony for the Freedom Tower on July 4, 2004. *Photo by Lester Millman*

(Top) Freedom Tower cornerstone ceremony on July 4, 2004. *Photo by Lester Millman*

(Bottom) Chairman Gargano speaking in Brooklyn in 2004. *Photo by Lester Millman*

Transforming the lower Manhattan skyline with new construction at Battery Park City.

Photo by Lester Millman

lima beans for Mr. Lombardi at the grocery, and after carrying the broom and duster through our house while Frank carried the mop and pail, and after my sister Connie polished our brass mailbox so it glinted in the sun—after all that, it was off to our jobs as boilermen, stokers on my father's crew.

The truth is, he was hard. He didn't trust. He didn't go around saying, "Great job, Charlie. Nice work." His intentions were great, of course, but he fell short, as we all do. I'll give you an example, which still hurts seventy-five years later despite my proscription against holding on to old regrets. My father made it clear he thought I was not willing to work like my siblings, Frank and Connie. It wasn't the truth. I went to school and church, delivered papers, worked at Mr. Lombardi's store, and did lots of chores for my parents at home.

One winter night, my brother and I were shoveling the coal in a dark and scary basement with creepy shadows dancing on the stone walls and low ceiling. I shoveled coal in and opened the furnace door like I was supposed to, to give the fire enough oxygen to keep simmering but not so quickly that it would burn itself out.

The next morning while I slept "late" (6 a.m.), my father inspected my work. The fire had died. It was freezing down there. He was fuming. He came home and dragged me back there silently, our boots crunching the snow on the moonlit Fifth Street. Together we stood in the basement staring at the cold furnace in the dark. "What do you see, Charlie?"

"The furnace is out."

"Which means?"

"The people are cold."

"The people are cold. And what's *our* obligation?"

"To make sure that never happens?"

"And yet…?"

"Yes, Pop?"

Then he pulled the cord on the overhead bulb and pointed to the coal shovel, which was leaning against a steel pole. Pointed like a prosecuting attorney. "That's exactly where I left the shovel yesterday, Charlie. You didn't touch it last night."

But I had. Even as a boy, I had the habit—the need—to put everything back neatly the way it was, where I found it. It was a stunning accusation, because it went right to my honor. "You didn't go because you wanted to play in the snow." The injustice! "Maybe you don't like shoveling coal, sweeping the hallways, dusting the banisters. Maybe you're too good for that, a real American kid."

Does any kid *like* that kind of work? I did it because I had to, because he asked me—and by "asked," I mean that any son of an Italian father knows the consequences of even politely refusing one's father. Or "agreeing" and then shirking. I think maybe I spent the rest of my life trying to make sure my father—gone now for decades—would believe I hadn't left the shovel leaning on the pole, that I had kept the people warm, as he always did.

I kept trying to impress my father with a solid work ethic. Finally, one night in 1978, at the New York State Society of Professional Engineers ceremony honoring me as Professional Engineer of the Year—I was chair of the group's construction committee—he smiled at my accomplishments. And the seventy-four-year-old John Michael Gargano told me he was proud of me.

He died twenty years later.

My mother died in a horrible way that eerily echoed the loss of her firstborn child. She was carrying a giant pot of boiling pasta water from the "old" kitchen to the "new," and dropped it, burning a large portion of her body and eventually succumbing to infection.

Sealing the Deal

Trust and responsibility are at the heart of a great family. And any great team is like the traditional family back in the day. Everyone contributes according to ability. Everyone's contribution matters. And all this happens under the guidance of an able and responsible superior—the boss. A great leader ensures everyone pulls his or her weight, finds the right person for the right job, fosters leadership potential, and emphasizes the importance of each role in accomplishing the greater goal. I think of family and teamwork as the concrete in the foundation of any great social, political, or physical structure.

The last phase of laying the foundation in a building is pouring the concrete—sealing the deal, so to speak. We used 208,000 cubic yards of concrete in the WTC rebuild—enough to lay a sidewalk on every block of Brooklyn, all eight hundred miles. Watching the first drop of concrete on its way into the forms is always a moment when I hold my breath. There's something so final about poured concrete, even though I know full well that it's no more or less permanent than any other phase of construction. For example, on that Flatlands Avenue sewer job I ran in my twenties, one of my responsibilities was taking cylinders of the concrete they were pouring. We'd look at the concrete a week, two weeks, a month later to make sure it had achieved the strength it was designed for. If not, we'd bring out the jackhammers and do the whole thing over.

There's something magical in the process of concrete forming. It symbolizes the solidification of the team members' combined efforts and interactions. All the components of the mixture have to work in perfect proportion or you don't get concrete—you get expensive slush. The concrete tells the story about the planning that went into it.

In one of my first private sector jobs out of college—actually, a job I took while on vacation from another, but I'll tell that story later—I did surveying and field engineering work for a great contractor specializing in building schools. His name was John Oechslin. I didn't have an official title because I did a little bit of everything for him, mostly surveying. I was in charge of making sure the bulldozer guys upended only the barest minimum of trees for each job (because some things aren't reversible).

Once construction was underway, I inspected each new phase of the work, from making sure we'd poured the concrete evenly to ensuring we used each piece of equipment properly and optimally, especially when the company was charged by the hour for using it. I was a real stickler for details, and in fact my scrupulous attention to another contractor onsite, a guy named Mario Posillico, earned me the invitation to apply my skills at his company, and ultimately to become his partner. On my first day on the job with J.D. Posillico Inc., I was sent to inspect a defective concrete wall at a school in Plainview, Long Island, to give an opinion on how cracks might have developed in it and how it could be fixed. While I was there, Lee Harvey Oswald shot President Kennedy. Maybe it somehow remained in my memory; while that concrete wall was salvageable, not everything can be changed once the course is set in motion. Some things you do and say, you can't take back. You have to work around them forever.

But even concrete isn't "set in stone." One of the other similarities between concrete and families/teams is that concrete never fully finishes setting, or "curing," as it's known in the trade. It's kind of like glass in some respects—it dries, and it largely finishes setting within the first few moments of pouring, but technically the chemical reactions continue to occur for as long as it exists. With concrete, you need to check on its strength regularly—the same way every family and every team need to regularly check on their strength and bolster it whenever necessary. If you maintain the correct potency of concrete, there's a very good chance it's going to remain like that indefinitely, as long as it's properly situated and it isn't exposed to weathering. Well, technically everything is exposed to some weathering—that's the price you pay for doing business on Earth! But you can certainly reduce the amount of water it's exposed to, avoid situating it near temperature fluctuations, and anticipate potential stresses that might compromise its integrity. In the same way, you have to keep outside influences from unduly affecting or compromising the integrity of your family or team.

When you pour concrete, it releases an immense amount of heat. This can require actively cooling the concrete for a time, especially on larger jobs. The Hoover Dam, for instance, with its three and one-quarter *million* cubic yards of concrete, required just over 582 miles of one-inch steel pipes that circulated ice water through the concrete from a refrigeration plant. The first concrete for the dam was placed on June 6, 1933, when my mother was five months pregnant with me, and the cooling wasn't stopped until nearly two years later, in March 1935, when I was toddling.[9]

So much could go wrong with so much material playing so essential a role in keeping stuff upright. For example:

> *When finished, Salesforce Tower will be the tallest building in San Francisco. For now, it's a big hole in the ground. And at the bottom of that hole is a new, massive concrete slab—14 feet thick, spread nearly an acre in breadth, and ready to support 1,070 feet of glass, steel...and a lot more concrete.*
>
> *Pouring it all took more than 18 hours on a cloudy San Francisco Sunday. An armada of trucks delivered nearly 49 million pounds of concrete and brontosaurine pumps vomited it into the hole while a small army of rubber-booted workers scurried about, directing the flow. It was one of the biggest, longest concrete pours in history.*[10]

This kind of thing gets my heart pumping like a picture of Gina Lollobrigida in *Solomon and Sheba*. In only one case in which I acted as surveyor—it must've been around 1961 or '62—did I make a significant mistake. We were finishing construction on a junior high school—it was a time of great suburban development on Long Island and all around the outskirts of New York City. We'd already built the foundation and the walls, installed the steel beams, and bolted everything together. The last major part we needed to complete was the windows. As anyone who's ever done construction knows, everything has to fit together like a jigsaw puzzle, and while you get a little wiggle room with some of the components—you can adjust the steel to fit, for instance, depending on the bolts—you cannot wiggle glass. Glass doesn't wiggle. You can wait a few billion years for its "amorphous solidity" to sort of, well, melt. But who has that kind of time?

For just one of the windows—and to this day I don't know how it happened, as I conducted the surveys myself and supervised the construction—we ended up with a gap an inch and a half wide. That might sound pretty small in relation to an entire building, but for me that was a chasm (and it would have felt that way as well for any occupant of that room in winter). It wasn't hard to fix the mistake—the window company quickly retrofitted a piece to close the gap—but even still, that one experience made me redouble my efforts at becoming as exacting as I could be. By that stage of construction, it was difficult to tell whose mistake it was—it could've been from the survey, or how the foundation cooled, or how they built the walls, or a dozen other things. Regardless of what caused the problem, it was my responsibility for not having caught it, and I took that lesson with me for the rest of my career. I considered it a gift that I could learn it so soon and with so few repercussions. Over time, I've heard plenty of stories of projects gone haywire over missteps made during the laying of the foundation. Not every one of those stories has a happy ending. Similarly, sometimes there's a family or team member who just doesn't fit. Try as everyone might, it's like forcing a square plug into a round hole.

Tom, my older brother, was like that. He was seven years my senior and had gone back to New York with my father and uncle when they left us in Avellino. I didn't know him. And when we came back, I found him very...different. Off, somehow. He seemed angry, always fighting. He would never go to school. And when he was fourteen or fifteen, he'd spend all his time at the gym—he'd started to box—so he was never at home with us, never really part of the family. He was like that window. I could see myself reflected in him, but he didn't fit.

Tom had thirty-nine amateur fights. And then he turned pro in 1944 at age sixteen. His career got interrupted when he was drafted into the Marines. He won the Marines boxing championship, and for two years he went around the world putting on exhibitions. He boxed greats like Al Reid and Johnny Rinaldi. He made five appearances at Madison Square Garden and even boxed in Yankee Stadium. And then he went downhill. He married a girl he'd met at the age of twelve, and they had eight kids. My parents were never happy with his decisions. I gave him a job at Posillico, where he worked for years until he died of cancer.

I'm reminded of Tom whenever I watch workers pour concrete. It reminds me of the responsibility contained in that material, the path it sets in motion once the plans and dreams and big ideas become real and turn, technically, to stone. Setting the foundation can be the most auspicious moment in the long course of building something great, or it can go horribly awry. In the re-engineering of America's future, it's also important to double-check every one of those calculations. When you have a chance to build, make sure you do it right. Check the strength of your materials often. Mind your foundations and make them count.

Birds of a Feather

When I look back on some of my actions as a young man, I blanch. What was I thinking? But we all have to go a bit easier on ourselves— we were forming foundations on the shaky ground of adolescence. Anna was my first crush. A deep crush. She was crowned Miss High School (beginning my long association with beauty queens, including Phyllis George and Sophia Loren—what can I say?). I met her at sixteen when she asked me to escort her to her sorority dinner. I brought her a corsage, of course, and off we went with friends to the dance. Naturally, I could never let her know just how strongly I felt about her. I was Charles Angelo Gargano and I did not have to chase after girls. What an ass I was, like most young men always were and always will be—so much less mature than our female counterparts.

Anyway, I avoided her afterward even though another part of me wanted to get on my knees and beg her to be with me. She told her friends she didn't understand my behavior—of course she didn't. Meantime, I was pretty popular with the ladies (several of them compared me favorably with matinee idols at the time). Many years later, I heard from mutual friends that Anna had always desperately wanted to date me. I never forgot her and the feelings I had for her that I somehow couldn't cop to at the time. I hated that I hurt her. I hope my immaturity then made me a better man and a better partner today. I'm certainly much more likely to talk about my emotions now. It took me only seventy years—and it's still a work in progress.

But in those days, I spent many years dancing, never staying still long enough to consider grounding myself. My sister Connie and her girlfriends used to go dancing, and I would join them when I was about eleven or twelve. I was a happy kid, social, bouncy, and I guess they didn't think I was cramping their style too much. By the time I was in high school and through the beginning of college, I had become a half-decent dancer. The Latin music craze was in full swing. It was all Tito Puente, Desi Arnaz, and Celia Cruz. Our hangout was the Palladium Ballroom above the Rexall Drugs at the corner of 53rd and Broadway in Manhattan. In 1948, the Palladium went all Latin, and so did my dancing. Machito and Mario Bauzá. On the weekends, we would go dancing for hours and hours.

There was another place around the corner on 53rd and Seventh called The Latin Quarter, a three-story wedge-shaped building facing Times Square—with the famous Coca-Cola sign. That was like Manhattan's own Copacabana. Frank Sinatra, Ella Fitzgerald, Patti Page, Sophie Tucker, Mae West, Diahann Carroll, Milton Berle, the Andrews Sisters, Frankie Laine, and so on. One night our group went dancing as always. We were the best dancers there. Cha-cha-cha, rumba, samba, salsa, mambo, merengue, and of course the jive. Only certain girls were allowed into our circle—and no other boys could penetrate. We were the best. Anyway, one night, this guy kept slinking around the edges, trying to horn in on us. Good-looking guy. Early thirties. Watching. Envious. Slavering over our girls.

It was Marlon Brando.

Block Party

I've come a long way since such moments were the highlights of my accomplishments. I assessed and improved myself, clarified my values. Over the years, my priorities evolved. I provided real and durable benefits to my family and re-evaluated my personal and professional goals along the way.

But things changed drastically only after I opened that dance circle, expanded my sphere. Once you've cleaned your area and learned to keep it clean, you're ready to stake out your site. What are the parameters of what you want to build? What are your goals and mission based on your core values? How far can your immediate sphere of influence reach? What materials do you have to work with? How can you take advantage of the landscape and the natural resources around you? Who else can you bring into your circle as a helpmate?

You have to take risks to get anywhere in life. You miss 100 percent of the shots you *don't* take,[11] said the hockey Hall of Famer Wayne Gretzky.

I got married in November '58, and in '59 I had about two months of vacation time coming to me. I'd been on the Flatlands Avenue sewer job for the city for two years. I said to my new wife, Prudence, "You know, I think I'm going to get a job during vacation; we can use the money." We were living in Deer Park on Long Island, so I went over to Commack Road, looked around, and there was a school going up. I asked the superintendent in charge, "Who's the contractor here?" And he gave me the name of the general contractor, John Oechslin, Inc.

The contractor was doing about eight to ten elementary and junior high schools a year around western Suffolk County. He was the biggest player. So I went to see this John Oechslin. I asked if he needed any engineering or surveying done. He told me they had a surveying company out there to do all their work, Blydenburgh, a big company with all kinds of experience and sophisticated equipment. I said, "Well, I can do it just as well or better than those guys, and I'm sure it's going to be a lot cheaper if you let me do it."

"Eh, let me talk to my wife."

He came back and said, "You can start tomorrow."

So I underbid a giant surveying company.

I immediately set about making myself as useful as possible, increasing my value to the guys in the field. I'd get in my new fuel-efficient Opel and drive to each site, in various stages of construction, walk right up and say, "Hi. I'm Charles Gargano. I was just hired as an engineer for the company. Anything I can do for you, please let me know." They looked askance at first at this young guy. Then they remembered the name Gargano. My family was well known in the trade. They gave me a shot. On one site, maybe I could compute the amount of concrete they had to order to fill some oddly shaped space.

At another, I could measure a vertical angle with a little better precision. They were using all these old-fashioned tools—ancient transits and theodolites, and so on. I helped them get more modern surveying equipment—K & E was the best brand. I was obviously trying to be helpful to all the superintendents so they would give a good report about me to the boss.

After about a month, I went in to tell Oechslin I actually worked for the city and had just wanted to make some extra money for my new wife and me. He asked me what I was making. I was earning about $6,000 a year, which was pretty good back in '58, about $115 a week. He offered me $145, which would have brought me to over $7,500. But I said, "You know, I think I might go back to the city, because it's a very secure job. I get a lot of benefits."

"What's it gonna take?" I knew right then I'd rendered myself invaluable to the big guy.

I said, "Well, your carpenter superintendents are making about one hundred seventy-five dollars a week. I'm worth that, no?" He asked me to wait a day—so he could talk to his wife.

"You've got your one hundred seventy-five dollars."

That was $9,000 a year—a 50 percent increase over what I had made working for the city. It was twice as much as the average public teacher would make in those schools I was building. I immediately took a six-month leave of absence from the city of New York. I didn't quit outright—I wanted to hedge my bets.

And then on one of those Oechslin jobs, I met Mario Posillico. I proved myself not only able but extra challenging in terms of costing and bidding. He courted me for a year, kept upping his offers. In the end I asked for $225 a week—$12,000 a year, which was double what I'd made working for the city. He said, "That's more than my brother Dominic and I get!" I got it. Plus a promise of a 20 percent partnership within a few years. I had to wait for them to buy out an uncle, and along the way I leveraged the time it was taking them; I suggested I might break off and start my own company. They knew I'd take half their company with me—all my guys.

I was twenty-nine years old.

It wouldn't have happened had I not staked my claim, understood I could control to some extent the outcomes for myself and my family based upon my attitudes and actions. So, say you want to be a better mother, provide more for your children. You've been doing a lousy job so far because of this or that excuse, and you've always known deep down—or maybe not so deep—that you could do much better. Staking out your claim means establishing the borders of what *you own*, your responsibility and your reach, regardless of the terrain, the resources you've got, and those you need to bring to bear, assemble on site. You can do this by instinct and intuition. You know what you own, the real borders of your moral property. Your personal Ground Zero.

If necessary, think about the roads you'll need to take to get where you want to go, and stake those out, too. Maybe you need a certain degree, or to live or work in a particular place. There is no distance between where you are and where you want to go that you can't traverse if you plan properly, drive hard, and avoid as many potholes and unscheduled exits as you can.

Then it's time to expand your sphere, open the circle to see out and let others see in. Take a look at your neighborhood, your immediate community, or someplace you care about. What are its problems? Can you triage them and focus on the top three, not get bogged down in the weeds so much that you can't see the forest? I'll give you a few examples of small areas where I felt I could make a difference. For me, cleaning up my neighborhood was always both a literal and symbolic act.

Before Reagan tapped me for my first political appointment, I was still running Posillico when, on November 23, 1980, the Irpinia earthquake nearly leveled my ancestral home, Sant'Angelo dei Lombardi, Avellino, in Southern Italy. There were ten thousand casualties—twenty-seven hundred dead in the region and three hundred in Sant'Angelo alone. Among them, twenty-seven children in a single orphanage died. More than a quarter-million people lost their homes. Eighty percent of the town where I spent my first four years was obliterated.

I had to do something. So I gathered a diverse coalition of politicians, celebrities, and industry giants, from whom I raised a lot of money for disaster relief aid. Then I traveled to Italy and landed in a helicopter with a small contingent of them: Republican New York Senator Al D'Amato; former Republican Massachusetts Governor John Volpe; my old friend from Brooklyn, Anthony Bevilacqua, who was archbishop of Philadelphia at the time; and Maria Pia Fanfani, the wife of five-time Italian prime minister Amintore Fanfani. She'd organized two benefit concerts by Maestro Uto Ughi, one in Rome and one in New York, for the victims of the earthquake. She later become an official for the Red Cross and continued the humanitarian mission she began resisting Nazis in the 1930s—an incredible woman.

Together, our ragtag band managed to get into those mountains a week after the disaster, loaded for bear with much-needed money. We brought hundreds of thousands of private American donation dollars— that was Mara Pia's specialty—to support the families of the victims, to assist widows and orphans. I gave $500 per family—a pittance in the big picture, but I hoped it would help. On subsequent trips, I brought more private and corporate American aid dollars to build prefabricated schools, a ball field, a nursing home, a computer learning center, and other essentials. But we were so successful that first week that Pope John Paul II asked to personally express his gratitude. One of the highlights of my life was sitting in the pope's private living room. There I was at the papal apartments in the Apostolic Palace in Vatican City at seven o'clock on a Saturday evening as the Holy Father thanked and blessed us under a medieval painting of the risen Christ. It doesn't get any more humbling than that.

It was a good thing foreign groups like mine helped out. The official relief aid from Rome was a debacle, with billions upon billions of *lire* misrouted to a new class of regional millionaires, misappropriated and ending up in the pockets of politicians, and misspent on pure pork[12] while the people froze their asses off in makeshift housing, some of which is still there in those cold mountains. It's heartbreaking to see my old neighborhood like that.

Harlem Nights

Sometimes a 6.9 quake wrecks a village in seventy seconds flat. Sometimes a single, rogue, shock-and-awe attack brings down a complex of skyscrapers in a matter of hours. And sometimes social and economic forces slowly, methodically erode once majestic neighborhoods through drugs, crime, and abandonment.

Look no farther than Harlem. The year I was born, a year after the Twenty-First Amendment abolished Prohibition, there was no place on earth like Harlem. Five hundred speakeasies suddenly came out of the closet (there were upward of a hundred thousand throughout the city[13]). The sound of jazz clarinets and drums filled the night air. Poetry flowed. But the trouble had already begun. The Depression hit Harlem hard. One in four of the residents—70 percent black—was out of work. Riots would start the following year—and not ever really stop.

The blight in Harlem had broken my heart since the first time I saw the neighborhood. When I was finally in a position to do something to help, once I became an official in city and state politics, I undertook a massive renovation project, piece by contentious piece. As chairman and CEO of the Empire State Development Corporation (ESDC), I automatically chaired many subsidiary projects, such as the Harlem Redevelopment Corporation. This included revitalization of the oft neglected nine-mile stretch of New York's "forgotten waterway," the Harlem River; and of Harlem's iconic "Main Street," the 125th Street corridor. There were fits and starts, partners that came and went. The economy was our biggest enemy. But over the years, we brought in huge anchors—department stores, grocery chains, banks, corporate complexes, car dealerships, pharmacies, cultural institutions, condos—and provided funds to renovate mom-and-pop shops. At the center of the new Harlem is the restored Apollo and Victoria Theater complex on West 125th and Lexington, a $50 million-plus mega-renovation. The dream, which I shared with many locals and city officials, was to create a new Harlem Renaissance. It was to be the economic equivalent of the artistic boom that occurred in the 1920s and was associated with the

"New Negro Movement"—involving novels, poetry, film, the visual arts, and, of course, architecture.

Sure enough, the ESDC's moves spurred new gentrification, private renovations, and other positive developments—and a good deal of controversy. We didn't go as far as I would have liked. After seven years and five extensions, our project's $48 million was sitting there waiting for the likes of Democratic New York Representative Charlie Rangel and his "Gang of Four" to get their acts together. Charlie was too busy collecting rent checks from his villa in the Dominican Republic to give a shit about his own neighborhood. We pulled the plug on a major development. I told the *New York Times* I recommended to Pataki that we abandon the Harlem International Trade Center in favor of finding a more viable host.[14]

But I look today at Malcolm X Boulevard and its environs, the retail stores above the Adam Clayton Powell Jr. State Office Building, and the former Hotel Theresa—and I'm struck by the positive effects of our parcel-by-parcel reconstruction. Is it perfect? No way. Is it better than it was and would have been? Absolutely.

I understand the controversy. Tensions were bound to be high as we used the state's authority to condemn buildings, as corporate money flowed and civic and community groups felt power shift away from their traditional home base. Temporarily upsetting the sense of local citizen empowerment was necessary to finally get something done up there, especially when Rangel and his cronies had such a liberal stranglehold. Some merchants had to be displaced in favor of giving residents more and better commercial options. I'm sensitive to the racial and ethnic challenges inherent in reimaging a community with the kind of history Harlem has—there's bound to be strife. Growing pains always accompany evolution. But growth is inevitable. Go with it.

"If I Can Make It There…"

And then there was 42nd Street, speaking of controversy. Some people say they're of two minds about the job I did re-engineering it as head of the 42nd Street Redevelopment Corporation. They say they're of two minds about the "Disneyfication" of the once "authentically" grimy thoroughfare. These people miss what they think of as the gritty charm it used to have. I'm not among them. Good riddance, I say.

I remember the first time I saw the place. It was 1948. I was an impressionable fourteen-year-old interested in construction, architecture, and civil engineering. I was blown away by the throngs of people crisscrossing the boulevard, and fifty billion flickering lights between Broadway and Eighth Avenue, advertising all manner of entertainments, both pure and impure. All my prior experience had been limited to the lyrics of the Al Dubin and Harry Warren title song of the 1933 Busby Berkeley film 42nd Street.

It didn't disappoint. But back then, the streets were relatively clean, and the people—including many families—were enjoying themselves. I was thrilled to be there. Proud as I was of my Seventh Street boys and our Grand Army Plaza, that was Brooklyn—and this was the other "country" of Manhattan. Here stood many of the Big Apple's most seminal structures since at least the start of the twentieth century: Grand Central, the busiest railway terminal on the planet; Broadway, the heart of Manhattan's theater district; Times Square, the intersection of the world since its commission in 1811. And there within the space bordered by Eighth Avenue, 40th Street, Ninth Avenue, and 41st Street, was the site of the future Port Authority bus station, Republican mayor Fiorello La Guardia's baby, which would consolidate eight far-flung stations across the city.

The mayor had died the year before of pancreatic cancer and possibly a broken heart—owing to Democrat William O'Dwyer finally getting his office. During the next two years, nine thousand tons of structural steel and more than two million bricks would be used to build one of the greatest transportation facilities in the world, all under the aegis of the agency that would become my future home: the Port Authority, which had been around since it was created by an Act of Congress in 1921.

The area had a storied history. 42nd Street and Broadway, the southeast corner of Times Square, was the eastern terminus of the first hard-surface, improved road that was to cross 3,400 miles of the U.S., beginning in 1913. They called it the Lincoln Highway. The idea of it eventually culminated in the formation of the Federal Highway Administration (FHWA) and the Interstate Highway System. Between First and Second Avenues on the East Side, the Hospital for the Ruptured and Crippled (now the Hospital for Special Surgery) tended to wounded veterans of the Civil War. The United Nations, the New York Public Library, Bank of America, and the Ford Foundation all grace its pavement. Yet it was always "naughty" and "gaudy" like the song in *42nd Street* says. In a 1900 photograph of "Longacre Square" (it didn't become Times Square until 1904, the year my father was born), you can see behind the horses and carriages on the unpaved street a billboard advertising "Burlesque: Ballet & Varieties." Varieties my eye!

Seventy-five years later, the infection of such enterprises of ill repute had spread to a leprous degree. The whole area had fallen into disrepair and moral turpitude, to say the least. Grindhouse theaters. Shady strip bars. Destitution, prostitution, pollution. By the '80s and '90s, when I first got involved, it was at a crossroads (if you'll pardon the pun). 42nd Street, especially west of Broadway, was a corridor of sleaze ending in the dreary Hudson Yards—the former docks and slaughterhouse district. Yes, Studio 54 basked in some of its glamour in the '70s—though its underbelly was certainly seedy. And, yes, Wall Street surged in a temporary flourish in the '80s—but it would be short-lived, given all manner of corruption and greed. In short, all of New York was suffering a crime wave, crack, AIDS, disintegrating infrastructure, and homelessness. It was a mess, and the epicenter of it was 42nd Street.

No self-respecting parent from Connecticut or Westchester County was comfortable sending a child to a Broadway show on a bus that unloaded its precious cargo in the cesspit of "Port Authority," that metropolis of skulduggery, which had once served as the gleaming jewel of the city. The surge of humanity that had moved into the city for two hundred years was abruptly reversing itself. An economic exodus left the dregs like a new bum's Bowery. On top of all that, streams of corporations, banks, and brokerage houses were fleeing the crumbling city.

It was time to act.

So, ten years before the tragic events that turned Lower Manhattan into a war zone, I saw 42nd Street as the neighborhood most in need of resurrection and revival. I adopted it. Instead of the grim urban nightmare it had become, I saw endless opportunities. I told Pataki soon after he appointed me, "Let's rebuild 42nd Street."

"They've been talking about that for years."

"Exactly."

"All right. Let's do it," he said.

And I did it. I'm using the word "I" here, though Pataki backed me 100 percent. He said, "Tell me what you need."

"I need Rebecca Robertson," I said. The Democratic president of the 42nd Street Development Project, the state agency in charge of the block, would become a critical partner. We'd worked well together in the past, and I knew our collaboration would bear fruit.

Administrations before had tried and failed. Dinkins and Cuomo couldn't get it done. It got hobbled by lawsuits, interagency squabbling, and a recession or three. The agency I headed, the ESDC, was responsible for developing more projects than any agency under any other governor in state history. I knew I could get the 42nd Street job done. I told the governor, "If I make a deal with Disney, and they agree to come on the street, that's going to help us revitalize the whole area. I can get them to take over the New Amsterdam theater, and that development will attract all kinds of major companies to invest. And what this will do for the West Side will be tremendous."

And it happened exactly as I said. Some nonbinding work had been done and undone, then redone several times over in advance—but by then whatever Disney negotiations had taken place lay dead in the water, like Dumbo meets Murder, Inc. The project had gasped and drowned for fourteen years. Much of the deal hinged on Disney's understandable concern about the seamier elements in the neighborhood; those shady constituents needed to be vacated before Disney could risk its brand being involved. I can tell you, it was no small feat getting those squalid porn dens to go—lots of lawsuits, some payouts, and so on. Ah, the power of the "condemned" sign. But we did it.

For five years after I started building the street, a slew of others slinked out of the shadows to steal credit. This included Rudy Giuliani (no surprise), who fought with David Dinkins in the press over credit, though it didn't belong to either of them. "The deal that was made to bring Disney to Times Square was made in my office, and announced by me,"[15] Giuliani told the *New York Times*. Funny old world—for some reason those guys signed my name and George Pataki's. Look, it's not about credit for the sake of credit, but the truth will out. The *Times* met with me several times, supporting my progress at every stage.

Ann Tighe, one of many who worked with me on the deal that brought Condé Nast to 4 Times Square, said, "More work has been done in the state under [Gargano's] watch than since the days of Robert Moses,"[16] another influential and controversial figure in the city's history. That statement, indeed, is one I regard as the ultimate compliment— despite occasional missteps and controversy, the legendary Robert Moses has always been a hero and role model to me.

With the help of my staff and Robertson, I made the deal with the Walt Disney Company in April '95, and we signed it in Pataki's office. For $36 million, mostly in state and city loans at a congenial 3 percent interest rate, Mickey Mouse could handle the investment without having to seek supplemental income down on the docks. Disney inherited "America's Main Street." In the absence of the peep shows and architectural abscesses that the Port Authority had been able to roust under eminent domain, Michael Eisner and his brand got an empty canvas, a blank slate to repaint into the showplace of old Broadway, where Disney could stage signature productions like *Beauty and the Beast* and *The Lion King*.

They'd failed in Paris. They'd gotten booted out of Virginia. Their efforts in Japan had cost nearly double what they'd projected. They were about to embark on their own "town," Celebration, Florida, a master-planned community that would wind up a splotch on their image. The home builders were underqualified and over-rushed; buildings started to fall apart. Real foreclosures swamped the perfect town, even caused a suicide. Wife-swapping, vandalism, even a brutal murder—the pixie dust on which Celebration had been founded suddenly turned to shit. "[I]t rankles us to see the 'American Dream' parroted so perfectly by a conglomerate that is seen, more and more, as a merchant of camp and artificiality."[17]

So New York's 42nd Street was a major risk in Disney's calculus—and a risk, as well, for New York to include it as partner. I applauded Disney's courage and foresight. There were those who feared that 42nd Street would become Celebration North—the whole thing a ridiculous, patently commercial, chintzy, fake-utopian sham. I vowed to do whatever the city and state could to make sure that didn't happen.

And it's not what happened. The area took off. Disney's New Amsterdam would become the centerpiece of a $1.8 billion complex of entertainment that would draw millions of locals and tourists and bring billions into the neighborhood. The retail entertainment strip we built on that street—theaters, cinemas, stores, restaurants, game parlors, offices, and hotels—rivals the best developments anywhere. Across from the New Amsterdam, the New Victory reopened as a nonprofit youth theater. A steel-and-glass brew pub opened at 42nd and Seventh, followed by a deluge of new dining and retail spaces. The Disney Store opened in time for Christmas 1996.

The whole thing was a great example of public-private partnerships (PPPs or P3s) in action, and more on those later. Fabled New York development companies like Tishman, Durst, and Ratner got involved—all would later become major stakeholders in the WTC rebuild. We brought in Madame Tussauds and an AMC theater, and renovated the Empire Theater—a Beaux Arts magnum opus—into the multiplex's grand lobby. Durst helped bring the Condé Nast magazine empire to 4 Times Square—later we'd bring them on as the first tenants in 1 WTC.

Our efforts on 42nd Street kicked off further progress, enhancing the quality and expanding the boundaries of the neighborhood. If you look at the improvements now on the West Side of Manhattan—the best view of which is from 1 WTC's observation deck or the Amenities Floor (64)—it's hard not to understand the scope of the improvements that all had their genesis in that "Disneyfication" some didn't favor. Together we transformed "the Deuce" (the once infamous span of 42nd Street from Times Square to Eighth Avenue) from a porn-addled, rat-plagued, drug-infested alley of inequity back into America's beloved "Main Street."

Regarding the resurgence of Times Square, its shimmering skyscrapers, grand hotels, and recreation venues, President Trump said of me—don't forget the New York accent—"What he's done on 42nd Street is incredible. I've watched what people before him did, which was nothing. He made it happen."[18] Let's give credit where it's due. Donald Trump knew what economic development was all about. And maybe it wasn't always *all* about him.

It's Off to Work We Go

After the personal space you occupy, along with your family and your neighborhood, comes your career. Family dynamics, especially how you handle chores and other responsibilities, set you up for professional success or failure. Are you a hustler? How's your stamina? Did you invent clever and creative ways to avoid certain jobs? Did you do everything in your power to get the lawnmower started just to avoid your dad's doing it in front of you on the first pull? Did you figure out you could put off cleaning your room until your mother finally did it? Are you modeling for the people both below you and above you what it means to provide real value? Did you use the coal shovel after all?

If you're not absolutely positive why you're on the payroll at your job—I mean specifically *you* and not some random mook—then ask your boss. Make sure no one else can provide the exact brand of value that you do. Maybe it's trust. Maybe it's accuracy. Maybe it's the right combination of accounting and actuarial. Maybe it's the fact that you come in an hour early and never leave before the head of your company (and you therefore might have to make some reasonable adjustments to the other aspects of your life).

If you've still got rungs to climb at your job or in your career, are you the sort of person your company management wants to have around? This is not only about your work ethic and specific skill sets—it's about your personality. Are you a friendly, pleasant, measured person? Can you work as well on your own as you can on a team? Do you have an internal mechanism controlling when you speak up and when you shut up? Can you assertively argue a point when you know you're correct without becoming personally aggressive?

All these things you learn within your family, that extended apprenticeship of the real world. I've failed more than once in most of these areas—I regret my failings, but I always tried to learn from them.

On the day that communist rifleman took the life of President Kennedy on a Houston street, I was reporting for my first big job in the private sector. I got that job and became a partner four years later. In the twenty-three years I worked for Posillico, I paid my dues in the trenches of field operations at scores of big and highly complex civil engineering and construction jobs. I quickly ascended to vice president and general supervisor, eventually managing hundreds of employees, not because I'm some kind of genius and certainly not because I'm any kind of kiss-ass. My partners told me it was because I'm a hard worker, I'm honest, and I'm tough but fair.

To get a job done well, you have to put your head down and get to work. But don't forget to lift it up to listen to the ideas of others, even those who disagree with you. I spent most of my time in construction trailers with the team running day-to-day operations. They set up a CB radio frequency just to notify each other I was on my way. But I wasn't strong-arming anyone. I was listening and learning, guiding and providing leadership.

Think of Lincoln as the exemplar of this. Sometimes you don't have to look far for counsel. I never would have been able to grow a tiny eight-man operation into a construction giant were it not for the family members I brought in. They had a lot more experience than I had at the start.

Italy Would Not Have Required
Any Homework on My Part...

When Reagan appointed me ambassador to Trinidad and Tobago in 1988, I immediately embraced its people as my hosts and made it my home. I admit I had to study up. It was imperative to become something of a world expert on that place, its history, topography, and people. I did that by listening and learning. Its capital, Port of Spain, has about the same size and diversity of population as Washington, D.C. I started there. Then I went to the farms and savannahs, to the little villages under karst outcroppings and to the Creole peoples of the mangrove swamps. I wanted to be worthy of representing my superiors, the U.S. government, and ultimately you, the public.

The more I got to know the place and started fortifying its trade, energy, telecommunications, and investment sectors, the more it became clear that a good deal of the money on those islands passed through what's known euphemistically as a "hidden economy." You and I would call it the drug trade. Drug interdiction increasingly became my number-one priority. I wanted to focus the resources of my little part of the State Department there, especially as it would increase regional stability.

But ameliorating Trinidad and Tobago's drug problem was a monumental task, and one I felt the U.S. was not up to, at least not the way things were organized down there. The borders were porous. Local counternarcotic forces were...let's say less than robust. There were boatloads—literally—of drugs (cocaine, marijuana, and other types) flowing from Colombia through the Orinoco River into the Gulf of Paria and Claxton Bay, and then up to Piarco International Airport, from which traffickers smuggled their wares to Europe. There was also a little regional airstrip—Camden Base—south of the capital, much closer to the bay in Couva, which I suspect smugglers used as a forward base.

The U.S. Central Intelligence Agency (CIA) had two operatives working out of the embassy, covering the whole of the Caribbean. We needed them like a fish needs a bicycle. Typical spooks, they would schmooze at cocktail parties and presumably report back to Langley whatever "intelligence" they uncovered when not themselves under any influence. I never understood their value. Thanks to powerful senators like Al D'Amato, Bob Dole, and Paul Laxalt (Reagan's "First Friend" and three-time national campaign chair), I made some changes. I got the CIA station chief and his assistant shifted elsewhere in the Caribbean. And we brought in the Drug Enforcement Agency (DEA), which we really needed down there. Again, you start at home and then look outside your door to the rest of the neighborhood.

Remember that on your campaign to increase your value to your bosses, you probably have several different levels of people you work for. Who's your real boss? Your direct supervisor? Your customer? Your constituency? They all need to value you—and you, them.

"Liberty, Justice, and Freedom for All" Is a Neighborhood Perk

Once enough of us have modeled Ground Zero Values in our personal, family, community, and work lives, we'll become those thousand points of light to ignite the rest of the world again. We'll believe in inclusion, not exclusion. Exceptionalism does not require disrespecting others. We'll believe in opportunities. We'll believe in self-criticism for self-improvement. We'll encourage peaceful and productive dissent.

Contrast that with Vladimir Putin's country. For the past several years, Russia has intensified its persecution of those who balk at the unchecked influence of the state. The Russian government has blocked several websites and amended lists of "classified information," as well as prosecuted those who have voiced criticism of the country's foreign policies.[19] It has also adopted ever more repressive laws, and has come down viciously on anyone receiving funding from outside the country. This includes more than one hundred humanitarian and other nongovernmental organizations (NGOs). In other words, the Russia government doesn't want our aid. Many of these organizations, which are trying to help their own people, have been forced to close their doors rather than identify themselves as "foreign agents."

Our prison system is no prize. But incarcerated Russian citizens live in medieval conditions with little medical care, and there are reports of torture and other ill treatment. Russia has also continually "failed to respect the rights of asylum-seekers and refugees."[20]

Recently, a massive movement has gained traction among youth—"Putin's Generation"—who've never known a time before Putin. These young people are fed up with oligarchy, corruption, and massive income inequality—and they're not afraid to take to the streets in the millions to fight for change. Putin can crack down on that—and has, violently—but he cannot ultimately control the innovative ways these young people can communicate over social media. Eighty percent of Russians get all their news from state-run sources[21]—but that's bound to change. Try as some countries might, they can't really control the internet.

China's done a decent job of that so far, though. It has shown little indication of leaning toward democracy. This is a one-party state whose leaders typically remain in office for decades. There are no elections for national offices—only village-level voting to keep peace among citizens and allegedly combat local corruption—which is a joke. The government limits freedom of expression, as well as interaction with the outside world. The press and electronic media labor under severe censorship—see a pattern there? Those seeking political and government office in the highly competitive system must pass rigorous monopolitical ideological exams.[22] It's the stuff of nightmares.

Closer to home, Nicolás Maduro, president of Venezuela, claimed in 2015 that if the opposing party won power, he would institute a "civil-military union." He didn't mean the "military-industrial complex" that Eisenhower warned us about in 1961. He meant martial law. He threatened a massacre. Maduro has continually rejected the idea that his opposition could ever win power. Part of the reason is that public funds are frequently used without consequence to support Maduro's candidates. In addition, government supporters have been buying media resources and using them to promote the regime—again, like Russia and China. Unlike Hugo Chávez before him, however, Maduro's leadership has run a previously booming oil industry into the ground, taking its toll on the country's economy.[23] As I write this, the people are mutinying in the streets.

In fact, the results of oppression in all these places and those like them are violent clashes, deprivation, mass murder, and starvation. That's what happens when a government values its own power over the freedom of its people.

4

Lead and Liaise

"We will remember the fire and ash, the last phone calls, the funerals of the children."[1]

It felt like George W. Bush, in a speech delivered to the United Nations General Assembly at UN Headquarters on November 10, 2001, was talking directly to me and my Port Authority family. The Port Authority lost eighty-four of its own on September 11, including our executive director, Neil Levin. I worked pretty closely with Neil when he was counsel to Senator D'Amato. I was thrilled when Pataki suggested making Neil the executive director and crushed when we lost him. There were forty-seven civilian employees killed at the agency's Tower 1 headquarters, plus thirty-seven members of the Port Authority Police Department, either assigned there or responding to the attacks. Collapsing structures are not about the buildings but about the human toll they represent.

At Ground Zero, it wasn't just the Twin Towers and Tower 7 that collapsed. Many nearby buildings were destroyed as those towers fell. At the northeast corner of the site, the L-shaped, nine-story 5 WTC suffered a large fire and a fractional collapse of its steel structure when debris from the North Tower hit it—a section of fuselage from United 175 wound up on its partially collapsed roof.

Landing gear from American 111 slashed through the roof of the twenty-two-story Marriott World Trade Center (3 WTC), a thin, V-shaped tower between the Twin Towers, facing west. It crashed into an office next to the top-floor swimming pool. The hotel was at full capacity at the time of the attacks, with eight hundred-plus guests and a conference going on.

Some of the two hundred people who fell or were forced to jump from the upper floors of the Towers also hit its roof at between 125 and 200 miles per hour. There was nothing recognizable about them anymore. The whole building was cleaved in half by the collapse of the South Tower. Parts of the skeleton of the second tower smashed around the survivors clawing through the tangle toward each other across the severed lobby. Reinforced beams the Port Authority had installed after the 1993 WTC attack created a small survival zone in the lobby, even as people and debris rained onto the glass awning on West Street.

A thousand people escaped out of, or through, the Marriott that morning—it was linked to the North Tower. But at least fifty people, including eleven guests, two hotel staff, and forty-plus firefighters who were staging in the lobby, never made it out.[2] The National September 11 Museum now sits on the site.

South Plaza (4 WTC), a nine-story building housing an international bank and five trading floors (including the one featured in the Eddie Murphy film *Trading Places*), housed the entrance to the Mall at the World Trade Center at the concourse level. It was immediately east of Tower 2, whose collapse wiped the whole building out.

At the corner of Vesey and West in the northeast quadrant of the site sat 6 WTC. At eight stories, it was the shortest building on the site. It housed the U.S. Customs office and was demolished beyond repair by the collapse of Tower 1, which had loomed immediately to its north.

Across West Street, built on the landfill excavated from the original Twin Towers' construction, six skyscrapers formed the World Financial Center (WFC), which had been connected by the Vesey Street Bridge over the West Side Highway between 1 WTC and 6 WTC. A colossal rod of steel penetrated the east side of 3 WFC—the fifty-one-story American

Express Tower—and massive debris marred the lobby and lower floors. It was close to collapsing.

The Vesey Street Bridge, informally known as the North Bridge, which linked the upper levels of the Winter Garden Atrium, was destroyed by the collapse of 1 WTC. The original Liberty Street Bridge was called the South Bridge—a companion to the North Bridge with a similar design. It sustained significant damage during the 9/11 attacks but was not entirely destroyed. It was repaired, extended, and reopened in April 2002.[3]

Between Liberty and Cedar Streets directly to the south of the site, the gorgeous St. Nicholas Greek Orthodox Church was completely buried by the collapse of the South Tower—it was the only religious structure destroyed in the terrorist attack. Before the collapse, parishioners reported seeing airplane parts from United 175 and bodies on its grounds. Santiago Calatrava, the Spanish architect who built the Transportation Hub, is halfway done reconstructing the church.

Just south of the church, the north face of the spectacular twenty-three-story Gothic Revival building at 90 West Street was only three hundred feet from the South Tower collapse. Office workers were killed there, and a days-long firestorm gutted the whole building. Butted against the 90 West building, 130 Cedar Street and its neighbors suffered ruinous fires, too.

Right behind those to the east on Liberty Street, and equidistant from the South Tower, was Bankers Trust. You might know it better as Deutsche Bank, and you probably remember the huge black shroud workers hung from it. The collapsing South Tower gouged a twenty-four-story laceration into the façade of Deutsche Bank, spilling its steel and concrete guts like a giant iron-clad soldier. The thirty-nine-story building was a total loss after that damage. Water, toxic waste, and mold did its share of further damage. For years, recovery groups continued to find human remains in the ballast gravel on its roof. The site is off the original sixteen acres, so its use is less restricted than that of others. The Port Authority prepped the ground for construction in 2013.

Since 2014, the Port Authority Police Department has used the space for vehicle parking, but that's temporary. There have been a couple of plans for redevelopment, including transforming the walkway in front of 5 WTC into a supermarket. Another plan would have had JPMorgan Chase build a forty-two-story tower. Right now, the site is awaiting a request-for-proposals period issued from the Port Authority, so everything's on the table.

Even as I knew it would take years to clean up and rebuild all this, I was bolstered from day one by the sight of the workers in those bucket brigades. In the heady and emergent first days, they included one thousand ironworkers from across the nation, and thousands of doctors, EMTs and paramedics, Marines, sailors, soldiers, NYPD officers, New York state troopers, FEMA workers, Coast Guard folks, National Guardsmen, and others. Each inspired the next, and all conspired to keep working despite the conditions until the cleanup was compete in May 2003, almost two years after the terrorist attacks. They motivated the nation to remember what we stood for: "…and the people who knocked these buildings down will hear all of us soon."[4]

In this case, surveying and taking stock meant seeing and feeling the enormous support, the team effort, like soldiers must experience in the trenches. It meant knowing I was not alone. I can't think of a single leader involved in the cleanup and reconstruction who didn't take this kind of personal responsibility.

But it also meant calculating the enormous effort it would take to get the job done. It never occurred to anyone to just give up. Necessity outweighed impossibility. It took the combined effort of tens of thousands of people—and hundreds of thousands of financial supporters—to get the job done, but get it done we did.

I wish I'd never had to learn the lesson, but the experience taught us all that we can conquer any challenge, clean up any pile we've made in our own lives or others have made for us.

As long as we have a team and a team leader, there is no problem we can't solve.

The 1,776-Foot Perspective

To accomplish that lofty goal, leaders have to look beyond the reactionary, immediate, short-term problems and look to the more important, long-term, visionary solutions. This isn't easy for our leaders to do because it isn't easy for humans to do. Never has been. Way back when, our ancestors were constantly worried about the saber-toothed tiger biting someone's face off or snatching a baby. There was neither the time nor the mental and physical resources to build tools to keep the tigers out of the camp in the first place. This is still a challenge today.

In your own life, how far can you see ahead? How far do you plan ahead? I wound up satisfied with where my career plateaued. I got there by consistent values and the missions that arose out of those. I decided based not on personal profit, promotion, or public praise but on my values. Goals and plans come from missions. And missions come from values. Goals have deadlines. Plans involve the passage of time. Even missions can change over the course of time.

But values are timeless.

Which is good because some of our missions might take more than our own lifetime to accomplish—they're transgenerational and they require that kind of "longpath" thinking,[5] to use the parlance of futurist Ari Wallach. Issues like long-term environmental sustainability, climate stabilization, racial harmony, global nuclear disarmament, and hunger.

If all you do with your time and mind is put out fires, you'll never get anywhere, not even to your valueless goals. It's the same whether you've caused the fires or someone else has caused them. You'll never even be able to think about those important values you'd like to act on. Even if you're able to keep your head above water, you'll still drown in regret come the end of your life.

Instead, look below. Pick one value. Start today. Start out small. Start at home. But start to act in service of the value instead of reacting to immediate circumstances. Don't think of it as a chore or a problem to be solved but as an opportunity to master your own life. True leadership is not even about problem-solving per se. It's about building on values.

The following list of values is inspired by the MasonLeads program at George Mason University;[6] it's a good example of the kind of Ground Zero Values that built our country and can sustain it though an uncertain future.

- Respect, empathy, and compassion for self and others. Think Lincoln.
- Service, contribution, and making a difference. Some call that legacy. Think Teddy Roosevelt.
- Integrity, moral courage, trustworthiness. Think George Washington, Gandhi.
- Transparency, consistency, congruency, authenticity. Think Margaret Thatcher, Nelson Mandela.
- Mettle, taking a stand, acting boldly for good. Some call that grit. Think Galileo, Socrates.
- Humility, ability to see your limitations, openness to other perspectives. Think Benjamin Franklin, Dwight D. Eisenhower.
- Wisdom, reason, the 1,776-foot perspective, long-term thinking. Think Queen Victoria, Golda Meir.

I have never met an effective leader who didn't possess all these strengths, who wasn't able to call on one or more of those values in his or her every action affecting themselves or others. Even those I knew well and with whom I strongly disagreed on key issues—people like Ed Koch, Mario Cuomo, and Democratic New York Senator Patrick Moynihan—I respected because we shared some important core values, such as intelligence, service, kindness, and diligence. Though their beliefs and actions based on those values sometimes differed from my own, I couldn't help but respect them for who they were.

So start with yourself in your personal life and interactions, then in your family, then in your local community and workplace, and finally, in the context of your Americanism, in service of one or more of our Big Ten punch-list items. Infuse your personal, familial, business, and political actions with values. Finally, encourage them in others—especially those with whom you disagree—and help bring our country that much closer to achieving its ideal. This is about much more than building literal bridges across literal rivers, though those are important, too. The idea I'm talking about is people of good faith, meant not to toe the line but to actively disagree, as long as they never lose sight of their duty as citizens of the greatest country the world has ever known.

If we simply return to the values of family, faith, and nation, we'll be on our way. In fact, *that*—not a manufactured home in Ronkonkoma or that 75-inch TV—is the essence of the American Dream.

Look no further than his words and deeds to see what our current president values. Think about it. He values self-aggrandizement, his family's fortune, vengeance, and authoritarian rule. Now contrast Trump's values to those of George Washington, who sacrificed personal gain for public service, who refused to be king, who self-limited his term in power. Or those of Lincoln, who made the ultimate sacrifice to unite us. Or of Eisenhower, whose strong leadership, humility, and personal responsibility took our nation from the brink into the modern era. Or of Harry S. Truman, who continued Lincoln's legacy of civil rights action by, for example, integrating the U.S. military.

A Thousand Points of Light and a Naked Emperor

It's not like any of us has to tackle the nation's problems alone. Great leaders understand the necessity of collaboration. They put teams together. A great team cannot survive with a bunch of yes-men. Whether you believe as I do that God created the world, or whether you focus on a cosmic big bang, the universe started as one unified thing, then shattered—divided itself—into innumerable pieces, permutations, kinds.

It's foolhardy to dismiss real diversity as mere liberal political correctness. No, diversity is the nature of the universe. God and physics lead us to the same irresistible conclusion: taken together, the world is one. But that includes all species, all nations, all religions, all sides. The more divergent perspectives you get, the better your overall view. We're a nation of communities, says my friend George H. W. Bush. "This is America…a brilliant diversity spread like stars, like a thousand points of light in a broad and peaceful sky."[7] Well, I don't know about the peaceful part, but I otherwise agree wholeheartedly.

The imperfect project our ancestors started in this nation is the most noble experiment ever undertaken in human history. It has the greatest potential for positively changing the world on micro and macro levels. It cannot succeed—especially given the exigencies of this latest re-engineering phase—in an environment where people focus on their differences and not the ties that bind us.

Why else, as our nation lay mortally wounded, would our greatest leader, Abraham Lincoln, have chosen his bitterest rivals for top cabinet positions? New York Senator William H. Seward, who ran a vitriolic campaign against him, became Lincoln's secretary of state. Ohio Governor Salmon P. Chase became secretary of the treasury, and Missouri's distinguished elder statesman Edward Bates became attorney general. Each of these men was better known, better educated, of higher birth, and more experienced in public life than Lincoln. They also disagreed strongly with his political opinions, especially as they concerned abolition. Yet he chose them because he knew the team of rivals would immeasurably help him lead.

He first won over Seward, charming and flattering the former senator with disarming humility, genuine grace and compassion, and razor-sharp wit. Within the first year of Lincoln's presidency, Seward became Lincoln's closest friend and adviser in the administration. Bates and Chase soon followed suit, gradually allowing Lincoln to repair their injured feelings. Key to this transformation was Lincoln's willingness to assume responsibility for the failures of subordinates, to share credit with ease, and to learn from his mistakes. Lincoln had "an…unparalleled ability to keep his governing coalition intact…" He kept the strong egos of the men in his cabinet in check. And this "suggests that in the hands of a truly great politician the qualities we generally associate with decency and morality—kindness, sensitivity, compassion, honesty, and empathy—can also be impressive political resources."[8]

Now juxtapose this with the psychological phenomenon called "groupthink." A team is assembled, or a group assembles itself along

some ideological, political, or corporate lines. To maintain their alle-
giance to, and the trust of, the group, they excessively quest for harmony
or conformity among themselves. At some stage, their common points
of view become ingrained in the group's identity. Or, maybe conversely,
each member blindly swallows the ideology of the leader hook, line,
and sinker. That's the way most humanities professors got sold their
bill of goods, and how and why they bait our most vulnerable students
with the same malarkey. But stick-in-the-mud righties are no better
than bleeding-heart lefties. Either way, you get unreasonable, irratio-
nal, dysfunctional—or even downright dangerous—decision-making
outcomes. Group members try to minimize conflict and reach a
consensus decision without critical evaluation of alternative viewpoints
by actively suppressing dissenting perspectives, and by isolating themselves from outside influences.[9]

Think about Pearl Harbor. No one on the Navy's whole team
predicted the attack, despite (in hindsight) numerous red flags. Or the
Bay of Pigs, when Kennedy's team refused to entertain the evidence in
front of them, nearly causing our country's nuclear annihilation.

William Dodd was a well-regarded historian, academic, and diplo-
mat who was ambassador to Germany in the mid-'30s, during the Great
Depression and the rise of Adolf Hitler. While in Europe, he was unable
to ignore the growing dangers of the changing political climate. Soon,
he could not deny the obvious signals. When he returned to the U.S.,
two years before his death, he conducted a twenty-city speaking tour in
which he warned his audiences about the threat of Nazism to democracy, a message he had continually expressed to FDR during much of his
tenure in Berlin—but which had fallen on deaf ears. Neither the president nor the State Department acted until it was almost too late.[10]

This happens everywhere, from the collapse of businesses like
Swissair to physical plane crashes, too. A copilot, either assuming the
pilot has things under control or unwilling to question the boss, simply
allows a plane to crash. Just speaking up—articulating a diverse point
of view—could save countless lives. That's why I feel so strongly about
what happened to my friend, Colin Powell, a Harlem-born four-star

general and W's secretary of state. Colin has thought of his 2003 speech to the UN, laying out the Bush administration's rationale for war in Iraq, as a blot on his record since the moment the world relied so heavily upon it to wage a certain kind of war. The speech set out to detail Iraq's weapons program, but as the intelligence would later confirm, there was no weapons-of-mass-destruction program in Iraq.

Colin had been set up by Vice President Dick Cheney and Defense Secretary Donald Rumsfeld. They were getting even with him for recommending to W's father, President George H. W. Bush, that the U.S. leave Iraq after we liberated Kuwait and defeated Iraq's Republican Guard.

More than thirteen years later, the speech continues to plague him— not just for what it got wrong but for the unintended consequences it might have set in motion. I'm not sure those potential consequences were quite such a mystery to Colin. Initially, he hoped that a speech before the UN would lead to a diplomatic solution. In the summer of 2002, Powell could see that the president was getting a lot of military briefings about how one would and should go into Iraq. But they didn't take us all the way through what might happen once we got there—not to mention how the hell we were going to get out.

Colin laid it out for George. He later recounted his private dinner conversation with the president on August 5, 2002, for a 2016 episode of PBS's *Frontline*. "Mister President," he said, "it isn't just a simple matter of going to Baghdad. I know how to do that. What happens after? You need to understand, if you take out a government, take out a regime, guess who becomes the government and regime and is responsible for the country? You are." Then he uttered the infamous phrase that later haunted Bush and his successor: "So if you break it, you own it." He wanted the president to understand that twenty-eight million Iraqis "will be standing there looking at us." He told me himself that he hadn't yet heard enough about the planning for such an eventuality.

Bush asked him what he should do. Powell said that the U.S. should try to avoid a war, and the way to do that was to go to the offended party, which was the UN. "Take this case to the United Nations and see if we cannot resolve the issue of weapons of mass destruction [WMDs] diplomatically," he told Bush. Bush agreed, and, with some reluctance, they decided together to take the matter to the UN to seek a resolution.

Colin was hoping that the Iraqis would turn over all the information needed to satisfy the U.S. that they did *not*, in fact, have WMDs or a major development program underway. Or that they would agree to give up whatever program they did have going. Neither was something Saddam Hussein was going to do. Why should he have done either?

So, in early January 2003, the president decided that military action was needed. Colin regretted his words to the UN forever after because, even at that point, he had his doubts. Now a coalition was beating its drums and tying its boots to go after Hussein as a violator.[11] He might have been a tyrant, a corrupt monster, a murderer of his own people, a supporter of terror—but he did not have an active WMD program.

Yet leaders across the world, including many international intelligence communities, believed the "evidence," and a brutal dictator got taken out. Maybe for the same proxy reason, O.J. Simpson got thirty-three years for a minor robbery in Las Vegas. The man had gotten away with murder. Somehow karma got him. Even Al Capone got done in, sent to Alcatraz not for the St. Valentine's Day Massacre but for tax evasion.

So it's not that Bush was unique in surrounding himself with mostly like-minded advisers. The current terminology for taking this emperor-has-no-clothes phenomenon too far is creating "the bubble." If you surround yourself with people intent on pleasing you rather than fulfilling your shared mission, you're screwed. And so is the project. This is especially true if you go out of your way to seek approval from others above seeking solutions to real problems beyond your ego.

Maybe there's no better example of groupthink than 9/11. The much-anticipated 2004 *9/11 Commission Report* cited our nation's "failure of imagination" as the overall antecedent to our defenselessness that day.[12]

How do you prevent failures of imagination? You surround yourself with savvy people. Ronald Reagan had the "Velvet Hammer," Secretary of State James Baker. Henry Ford was never afraid to put captains of industry in charge of projects, men who'd forgotten more about the technical aspects of his vision than Ford would ever know. Elon Musk puts ideas in cyberspace and makes them open-source so experts everywhere can get involved. They take his ideas and run with them.

Donald Trump has Jared Kushner, Don Jr., and Ivanka running our foreign and domestic affairs. That's terrifying.

The Will to Lead

I had the honor of working closely with George Bush Sr. What a good human being, a really good man. Purpose over ambition every time. Substance over flash. In 1992, I was invited to join Barbara Bush and the philanthropist Max Fisher to watch George Sr. debate Bill Clinton and that half-loon, half-mastermind Ross Perot. It was the first town hall-type debate among the three of them. Should have been a cinch, right? Bush had a 90 percent approval rating when he pulled the troops out of Kuwait and Iraq earlier that year. And then he went down, down, down.

Max, Barbara, and I were sitting up front. And we saw the president check his wristwatch for an inordinately long time. Barbara looked at me with wide eyes and raised one shoulder, horrified but trying to keep her composure. I froze. Holy Christmas! I knew it was over right then. George fumbled his answer to a woman's question about the candidates' personal experience with the rough economy. Ross Perot and Bill Clinton started kicking the stuffing out of him. And he's not that kinda guy. Like his son George W., he's not a natural-born fighter. Afterward, he admitted he couldn't wait for the damn debate to be over. Unfortunately, that came through loud and clear.

Forget president—I thought he was a wonderful human being, and still think he is. He held many titles in his lifetime, from head of the Republican Party to CIA director and on and on. At that point, I think he just didn't really want to be president anymore. His heart wasn't in it. He'd picked some good battles, fought hard for them, and now he was done.

It's the same reason I think my other friend George—Pataki—did poorly in the last presidential primary process. Even as I pledged my loyal support, I got the impression he was running for president because he actually wanted to change politics, change America for the better. Which is not the way you win the presidency. That takes much more guile and Machiavellian (ethical) compromise.

The courage and decisiveness Ronald Reagan displayed in standing up to the air traffic controllers is just one of the many reasons I adored him. I respected him. I got to know him well; I served him for six and a half years. I used to go to the White House a lot, and when I had to talk to him, I found he listened and spoke with me like a regular human, not a man obsessed with his own power. That homespun, chummy public side of him was not just a persona. It was the real him. That's not to say he wasn't a keen politician and a deep thinker. His greatest asset was "social intelligence." He really cared about people. He made you feel like you were the most important person in the room.

The Republican Party is missing that kind of model nowadays. Republican Speaker Paul Ryan of Wisconsin might be bright, but he doesn't have Reagan's personality, integrity, or leadership ability. The bottom line is, with Ronald Reagan, it wasn't about Ronald Reagan— it was about how, together, you and he could serve the people of this country.

Again, contrast that with Donald Trump. I worry about him because everything is always about him. I'll give you a good example. In 1998, he asked me to be a judge in the 48th annual Miss Universe pageant, at my old embassy in the Republic of Trinidad and Tobago. I thought about it for a while, and I spoke to Phyllis George, a former Miss America, Miss Texas, and First Lady of Kentucky, whom I used to date in those days. And she said, "Sure, let's go."

We went on Donald's plane. I paid, because I was in government, but he treated Phyllis. When we arrived, there were half a dozen camera crews from local TV on the tarmac. Two or three reporters came over to me and started filming. They were excited to see me. "Wow! You're back!" I'd been a pretty popular ambassador—my official title had been ambassador extraordinary and plenipotentiary—for three years, from 1988 to 1991.

Well, it was obvious that Donald absolutely hated the attention I was getting. He huddled with his staff, then sent one guy jogging over. When the poor fellow expressed his boss's displeasure, I said, "What does he want me to do? I didn't call them over. He and I got off the same plane together."

Then another staffer arrived. "Mister Trump would rather you didn't talk to the press at this time."

"He wants me to tell the media and the people to get the hell away from me?" The poor lackeys just looked at me as if to say, "Yeah. That's exactly what he wants you to do."

That's Donald Trump in a nutshell.

Innovate and Optimize

Good leaders consider new ways of doing things. They seek the competitive advantage. Sometimes a simple tweak can make all the difference.

Almost everything about the WTC rebuild was innovative, including the building techniques themselves; the sequencing, safety, and technology; and the environmental and aesthetic aspects. In many cases, some of the stuff we did had never been done before.

There have been many books written about innovations at Ground Zero, so let me take you back to the sewer instead. One of the real innovations that made J.D. Posillico Inc. a leader in our field was the use of steel sheeting to support deep sewer channels during excavation. For centuries in North America, since the early 1800s through Ed Norton's days in the New York City sewer system, wood sheeting was the go-to material to support the construction trench necessary for deep excavation in sewer lines. You'd site the project, ensure the placement didn't conflict with any other infrastructure (such as subway, electrical, gas, or telephone lines), dig out the area, then line up wooden sheets along the sides of the trench to prevent the walls from collapsing on your people and work.

From there, if it was a small enough sewer line, or not particularly deep, you'd lower in premade pipes and hook them up to each other. For larger lines, if you had to, you'd pour concrete into a mold, creating the pipes as you went along, as we had to on the Flatlands Avenue job. This was an okay method for the majority of jobs throughout New York City during my heyday in the 1960s and '70s—except it was bound to fail in the sands of Long Island.

There, you'd dig three feet below the surface and hit water. The water table is extremely high, as the surrounding water affects not just the shoreline but underground. Water runs the length of the island, which also possesses many bays, ponds, and rivers. It's an idyllic situation for most of the residents, of which I was one and still am. But for people working in heavy construction, it proved a real nightmare—and still does.

Essentially, to lay any major sewer lines—the kind that hook up entire streets or neighborhoods to a treatment plant, for instance—we had to dig down deep, in some cases twenty-five feet or more. And when we hit water at three feet, we needed to start pumping the water out of the trench—"dewatering"—to keep it from burying the work, ruining the equipment, and drowning the guys. In some of these jobs, we'd get half a dozen pumps at different depths all going at the same time just to give us the ability to continue digging. It was very expensive and time-consuming, and when it came time to lay the pipes, the wood we'd lay along the sides of the concrete molds would buckle or simply turn to mush.

There had to be a better way. Sure enough, we found it.

Steel was our answer. Steel sheeting gave us the support we needed to keep the water out of the concrete while it was setting, and keep the pipes from getting compromised by the surrounding groundwater. As simple as this solution sounds, it had never been done before. And though the initial investment was high—we expected to lose money on the job—it worked so well that it lowered the costs and sped up the process considerably.

We started winning job after job, outbidding every other contractor on Long Island and eventually all of New York, and we quickly revolutionized the way sewers were created in areas where they would have been prohibitively expensive—or impossible. Suddenly, new areas above ground could be developed because of the work we were doing far beneath the surface.

As unbelievable as it might sound, we achieved a kind of fame from the lowliest and least glamorous of construction types. We made the covers of industry magazines. Contractors, architects, and engineers came to watch us laying sewer lines. They needed to learn the new techniques fast; they were losing all their business to Posillico, because we had the lowest bids and the quickest turnaround time. And, boy, we really delivered. By the end of the 1960s, we had grown from a business of just eight guys to a company with three hundred employees. And by the time I left to work for Ronald Reagan in 1981, I was supervising a staff of eight hundred. Even today, that company's still growing.

The Toe Bone's Connected to the Heel Bone... Assembling the Mission-Based Crew

Teams work best when all the members do their little part, when they understand that if their cog in the big machine stops working, the whole kit and caboodle comes to a standstill. We need to take responsibility for our part of the American team—but we need to appreciate and respect all others, too. You might take your baggage handlers for granted, but if they're not good at their jobs, you're late, the airline loses money, the economy suffers. So why do they deserve less respect than the pilot or the airline exec?

The closer you get to the foundation of any major project, the more you find that valuable people are the concrete that holds it all together. One of the reasons it's so important that we begin the re-engineering right from the foundational infrastructure is because it's all so thoroughly interconnected—if any single resource falls apart, it also messes all the others up in a domino effect. For instance, a backed-up sewer can unduly strain a road, which can cause it to wear away or collapse, which in turn can prevent food and other goods from being transported. Every one of those resources depends on all the others, and if your ability to put food on the table can be compromised by a sewer built on hundred-year-old wooden planks, you'd better believe there's a crisis in your future if you don't act, and soon.

It's the same in the figurative sense for the people who form the chains that hold our values together. You get one bad apple and the rot can spread, collapsing the whole project. If it's the top apple—the leader—the rest of the barrel's going to have a tough time staying fresh.

Barrel of Monkeys

There is such a thing as *bad* collaboration. Let's look—and I really hate to—at Moynihan Station. The Port Authority was willing to put up money to develop a new Penn Station. It was beautiful. I could see it in my head, and I often walked through those gorgeous designs hanging on my wall and piled on my desk. President Bill Clinton had promised $60 million to $70 million of federal funding, so we thought we were golden.

But New York state answers to a Public Authorities Control Board. Your project needs unanimous approval from the governor, the senate majority leader, and the assembly speaker. In 2006, Sheldon Silver, as assembly speaker, nixed the deal; he was the sole naysayer. I strongly suspect that Silver's decision was influenced by certain bedfellows—specifically Patricia Lynch, former communications director for Silver and former lobbyist for James Dolan, owner of Madison Square Garden (MSG). Dolan felt that Moynihan Station would harm business for MSG, despite the city's offer to relocate and rebuild the arena less than a block away from its present site to avoid disruptions, *free of charge*. I can only assume that Lynch lobbied Silver between assignations to make sure the deal did not go through. It would certainly fit a pattern. "When Jim Dolan wanted to kill the West Side stadium plan because it could draw business away from his Madison Square Garden, he hired Lynch and it worked. The stadium idea was DOA."[13]

Let me explain what that last part means. At 840,000 square feet, the Javits Center is microscopic contrasted to the exhibition spaces at convention centers in comparable cities. (McCormick Place in Chicago, for example, boasts 2.6 million square feet.) During the first phase of the megaproject now known as the Hudson Yards development, which happened under my watch, I wanted to expand the Javits Center. In fact, I wanted to double its size. It was about a $1 billion expansion plan. Along with that, the ESDC proposed as a centerpiece of our expansion strategy the West Side Stadium, an Olympics-worthy eighty-five-thousand-seat arena for the New York Jets football franchise. It would have sat directly above the monstrosity the Metropolitan Transportation Authority (MTA) calls its West Side Yard. The public would have contributed $600 million, and the Jets another $800 million.[14]

My earlier proposal would have had the Jets and MSG jointly building that stadium. We even went to Tokyo to look at a similar retractable-roof "superdome." But then Dolan balked. He didn't want to share. I was telling him and the MSG people, "The Jets are only going to use it sixteen days a year! It's not like the Jets are going to the Super Bowl— this ain't 1968. You guys can run the facility the other three hundred fifty days." Again—the Dolan-Silver kibosh.

The media has since exposed the impropriety of Silver's and Lynch's relations, along with a slew of Silver's dirty dealings;[15] two juries convicted him of fraud, money laundering, and extortion, and he's currently trying to appeal his latest conviction by arguing, reminiscent of Clinton's impeachment hearings, for an alternative definition of "corruption"![16] You can't make this stuff up. Meanwhile, Jim Dolan has become famous for abysmal business decisions, including buying the failing Wiz electronics and entertainment chain—that cost him $250 million—and the Clearview Cinemas chain, which was also a major drain.[17]

Moynihan Station and the West Side Stadium are perfect examples of one rotten apple spoiling the bunch. You know, with Moynihan Station and the new MSG, the state had already spent $25 million to $30 million developing the project—planning, designing, and coordinating with the MTA, the post office (which owns the site), and dozens of businesses and institutions, both major and minor. But one person, Sheldon Silver, had a contrary agenda, and whether because of external or internal influence, he chose to scupper the entire project without further consideration. No leader in the world, aside from maybe Kim Jong-il or Tony Soprano (or his mother) could have altered the course of this situation with Silver on the team.

When it comes to re-engineering, collaboration is essential, and the wrong collaborators can harm everyone involved. To this day, the failure of those two projects remains my greatest professional disappointment and regret. Governor Pataki and I never should have promised Senator Moynihan to his face we'd name that station in honor of his service. It was something the senator had always wanted done.

"We Are All in the Gutter, but Some of Us Are Looking at the Stars"[18]

My father taught me the important similarities between families and work crews. He and my uncle Peter and several other relatives—including a skinny, hardworking kid named Charlie—worked for Hendrickson Brothers Asphalt & Paving in Valley Stream, New York. So when I eventually began working for J.D. Posillico, I brought on both my family and the Hendrickson Brothers family, as well as a few folks from the Oechslin days—people who knew the job, who had it in their blood. I watched several great leaders in all three of those places harness the strengths of those around them. They knew how to get the best from their crews by training them well, asking them questions about their expertise, then slotting them into the appropriate jobs to maximize their contributions to the overall project.

Assembling the right crew became crucial in the case of those deep sewer jobs we did in Suffolk and Nassau Counties from the late 1960s onward. This was after my father worked in concrete on the Verrazano-Narrows Bridge between Brooklyn and Staten Island, proud to find it named for our *paesano*, the first documented European to explore New York Harbor and the Hudson River five hundred years ago. My father, grandfather, uncle, and I were sometimes the first Italian Americans to explore stuff, too—deep under the surface.

Before I got involved in the re-engineering of Lower Manhattan, if I'd had to pick a single infrastructure project I feel most proud to have worked on, it would definitely have been the pipeline between Babylon and Bay Shore. This was along the South Shore of Long Island, a project I oversaw from 1975 to 1980. The city needed what's called a "trunk line," an immense, lengthy sewer pipe that would connect all the smaller pipelines to the treatment plant. It was an exceedingly difficult task. The pipeline had to go beneath the Robert Moses Causeway, a busy parkway that connects the Southern State Highway to Fire Island and two state parks. It also had to pass several canals that had their own high volume of marine traffic. All of this work had to be done beneath the ground—even beneath the water. We needed the best people in the world for this job—and we had them.

Like a bridge crew, a sewer construction crew consists of multiple interrelated operations. We had eight teams, all necessary to implement our proprietary new method of using steel sheeting to support the trenches. One team excavated the earth; another used pumps to keep water out of the trench; another implanted the steel sheeting; another laid the pipe; another extracted the steel sheets; another backfilled (replaced the earth in the trench where the pipe was laid); another prepared the ground for the next day's excavation; and the last coordinated all the work.

The potential for disaster was immense. We could have had caveins or wash-ins. The risks for all of us were very high, and we had to depend on the most exact calculations to make it all work. Remember, too, this was before the days of GPS or even cell phones. When work on one side of a canal began, we also had to work on the other side to make sure the two sides met precisely, like a mini transcontinental railroad. A few inches off would mean the difference between a successful pipeline and two giant holes leading nowhere—or exponentially worse, leading to the bottom of a canal. We couldn't disturb the traffic on the roads or waterways while we worked. You can't imagine the intensity of this task; it took years of careful planning and execution to pull this off, hundreds of workers all doing exactly what they needed to do in a vast sequence, everything proceeding like clockwork in what we call a construction "train."

I ran that job, making sure everyone understood and bought into the mission. We needed to lay one hundred feet of pipe per day. Using the steel sheeting method we'd devised and perfected, we succeeded. It made the front page of the *Engineering News-Record*, showing how much quicker sewer jobs could be done. Partly that was the technological innovation—but it couldn't have been done without the teamwork.

Every team member had to know what he was doing. It was the only way the operation could continue and move forward at a pace of one hundred feet a day. And they all knew it. It wasn't as though any task was less important than another. Each one was essential to move at that breakneck pace. If any of those eight operations faltered, it jammed up the whole project.

It worked magnificently. The project went off without a hitch, and it brought the South Shore up to modern plumbing standards. And do you know that to this day there isn't even a name for that pipeline? The project wasn't covered in the mainstream papers, and even now you can't google it to learn anything about it. It was a miracle of modern engineering and nobody even knows it exists, least of all the present-day residents of the South Shore. But if it hadn't been built, had been built badly, or somehow ceased to function? You'd better believe they'd know about that pipeline in a hurry! The fact is, that trunk line changed the lives of tens of thousands of people, improving conditions in homes and businesses in ways that most people can't even imagine today. We saved thousands of people money, time, and grief without even stopping the traffic. That's the kind of job that makes me proud of this country. I say the pipeline doesn't have a name. But I can think of hundreds of names of the men and women who made that work happen through teamwork, discipline, sweat, and expertise.

So the next time you see construction work on the street, just imagine what the workers are standing above, what's down in that trench. Chances are, if it's a modern city, it isn't just dirt. There might be up to two hundred feet of infrastructure beneath them—sewer lines;

air vents; telephone, power, and gas lines; and subways running every which way. And you'd better believe those workers know exactly where to dig and where they'd better not, or they can create millions or even billions of dollars in damage and risk the public welfare.

Imagine, then, what a renegade or misdirected construction worker could do by drilling a bit over here or blocking off something over there; the entire web beneath our feet could tangle up and incapacitate us. The potential for catastrophe with a "deconstruction worker" on the job is extremely high. I'm blessed to have worked with many fine people on my teams, all of whom were well-trained, motivated and, most of all, proud of the part they were playing in improving all our lives.

General Relativity

A great team requires dedication from each member—everyone's on board with the same goals and objectives, and everyone sees himself or herself as a vital, interlocking part of a larger project. All the team members know the tiniest, most insignificant-seeming cog can bring down a sophisticated machine. A smart leader will make sure all the people on the job not only know how to do their part but also understand how it fits into the whole. You could use the domino metaphor or the chess cliché. They both apply. So do the bottleneck analogy and the weakest-link truisms. You get real buy-in when all members of a team see themselves as *the* key to getting the job done. A good general fosters pride and morale during the battle and gives all the credit for the victory to the soldiers in the trenches instead of egotistically taking it all for himself or herself. Conversely, a good general takes full responsibility for failure and doesn't look farther down the line. No sense searching for a weak link when you failed to anchor the project in the first place.

George Pataki was just that kind of general. For the new WTC project, that man assembled a team of diverse talents and minds, articulated the mission, listened respectfully, led decisively, adapted when necessary—and ultimately got the job done. You don't see him trying to take credit for all that, though. He leaves that where it belongs: to the twenty-six thousand workers.

Seal Team Six

Obama, on the other hand, ate a lot of shit. Despite the romantic narrative, he was not a good listener, not a team player. As a result, he couldn't pull the trigger on much of value besides—and let's give credit where credit is due—allowing a Seal team to kill Osama bin Laden. Obama was also correct on two other counts: acting quickly on intelligence before the Pakistanis got wind of the impending operation Neptune Spear; and not releasing the photos of bin Laden's obliterated face. It's not the American way to "spike the football" that way, inflaming Middle Eastern ire with such a "trophy." If I were in charge, though, I would have put some conspiracies to rest by releasing DNA evidence from the corpse before we dumped it into the sea.

The actions of the seventy-nine heliborne commandos and their dog, Cairo, on the night of May 6, 2011, dramatically prove a point. All the team members shared the same objectives, and they handled their particular task quickly and correctly in a matter of thirty to ninety seconds. They were highly trained for all contingencies, and they employed work-arounds instantly for several issues, not the least of which was the crashing of one of their stealth Black Hawks and safeguarding the small knots of children clustered all over the compound.

With a great team you can do anything. The crews that worked the new WTC's fundaments—steel, concrete, and glass—were paramilitary in their discipline and mission-based operations. They kept to a grueling schedule of constructing two floors every two weeks. That took a lot of planning, setup, innovation, and motivation, most of it nonlinear—just like planning a military raid. "Everything—design, engineering, and construction—was developed concurrently and collaboratively."[19]

And that was for nine major buildings spread across a sixteen-acre holy land crisscrossed by subway and train lines, feet from the Hudson River, under intense security limitations, while several wars were being fought abroad. Everyone operated as a single unit. But one of the keys to our success was that although we all shared the same values and mission, we were all quite different in our approaches to problem-solving.

P.O.V.

What does diversity really mean in practice? A priest, a rabbi, and a handicapped horse walk into a bar? A deaf Mexican Jewish lesbian is going to be the linchpin of your team? Maybe. But it really means seeking diverse experiences and perspectives. Sometimes that coincides with diversity of gender, ethnicity, national background, political affiliation, and physical ability. But not necessarily. At ESDC, the top people were about 60 percent women—good, committed leaders in general, but also coming from a world that most male executives had no clue about. This benefits a team in two main ways. First, it gives the other leaders and team members a chance to hear all kinds of perceptions and evaluations they might not have otherwise heard. It helps them grow by understanding the impact of particular decisions on people who aren't exactly like themselves.

And second, it includes the communities represented by those diverse members; therefore they take part, at least representationally, in significant projects that require maximum participation. So a team will work optimally if its members have diverse points of view, a willingness to share (in the spirit of ultimate cooperation, not open hostility), and an approachable leader who encourages dissenting viewpoints. Such a leader will consequently understand the drawbacks and benefits of proposed methodologies, and maybe consider wholly innovative approaches.

Malcolm Forbes describes diversity as the art of thinking independently—together.[20] The Port Authority owned the site. Silverstein held the lease. The MTA owned and operated the subway lines. The state Department of Transportation (DOT) owned and operated West Street, State Highway Route 9A. Marriott International had the rights to operate more than half a million square feet of hotel space on the site. A company called Westfield America owned the rights to operate a further half-million square feet of retail space. There were insurance companies involved. Nearly ten million local residents all had their ideas. You bet that's inbuilt diversity.

Did you know that people from more than ninety nations were lost on 9/11? That every state in the union contributed in some way to the rebuilding—from Alaskan master plan communications coordination to Wyoming insulation; from Ohioan concrete admixtures to Oklahoman butterfly valves? The marble in the lobby of 1 WTC came from Italy. The elevators and escalators are German, their drives from South Korea. The oculus glass in the Transportation Hub is Austrian. It's a truly global world, and the "World" Trade Center became a truly global effort.

How do you manage all that? If you've got a leader who doesn't let ego get in the way and doesn't have some special narcissistic agenda, that leader will encourage dissention, listen to what the team says, weigh the options critically, then decide. If that leader also has the ability to get buy-in from the team thereafter—something Lincoln could do, and Reagan, and Pataki—your project is in good hands. The leader is, too. He or she isn't out there like some lone wolf on a solo journey. The whole pack is behind, around, or often in front of the lead wolf. If, say, the pack leader intends to ban a certain other breed from entering that pack's territory, at least the rest of the pack has been consulted and has advised, and everyone is prepared for exactly what to do once an alien encroachment is about to happen.

I was of two minds about Trump's Muslim ban—but at the very least, he ought to have gotten buy-in from the folks who would have had to put the ban into action. He acted prematurely and foolishly. I've seen both Cuomos behave this way in politics, trying to control every aspect of everything without getting support—they call it "mutual aid" in the fire service—from anyone else. When that happens, it's bound to fail.

Vox Populi

Very often, a team needs the input of the public at large. That's certainly true of a project as great as re-engineering America. That requires setting up mechanisms for receiving and processing the input, and communicating with the masses throughout the process.

The Northern Manhattan Parks 2030 Master Plan comes to mind. The New York City Department of Parks and Recreation developed the plan with substantial input from community residents, "friends of" groups, Manhattan Community Board 12, local partner organizations, park experts, elected officials, an advisory council, and its local (Inwood and Highbridge) operations teams. They did this by the book, with public meetings, workshops, site visits, interviews, interactive web platforms, and nonprofit and sister agency plan reviews. All that input helped shape the final recommendations from Parks and Rec.

Staffers researched the locations and number of recreational facilities, the health status of the community, as well as what social and physical obstacles most needed to be addressed. They interviewed neighborhood experts and park users. Professionals provided essential information. They asked the public to tell them what the Northern Manhattan Parks would, should, and could look like in twenty years if they could magically remove all barriers. Ultimately, the public envisioned a greener neighborhood more connected to parks; a linked system of scenic and safe pedestrian and bicycle paths; a healthy, natural, sustainable shoreline with public access; and more recreational facilities with restrooms. (Speaking as a former public official and a current eighty-plus-year-old, I say you can never have too many bathrooms.) The city even consulted with academics, using Princeton's Sociology Department's interactive website All Our Ideas to reach people who did not or could not attend meetings. Then they published an Interactive Public Input Map.[21] That's a model for public input, even if the city can't meet all those demands.

That all sounds wonderful. I can assure you it was not as easy and pleasant as it sounds, for either the organizers or the members of the public who intended for their input to be heard. Necessary as it is, public input can be a messy and contentious process. We organized many, many events involving the public in the process of reimagining, redesigning, and rebuilding the WTC site. Many more cropped up spontaneously and got our ear. The Partnership for New York City, which is the voice of the city's business community on legislation, regulation, and public issues that impact business and the economy, reported its findings to us. New York New Visions, a Coalition for the Rebuilding of Lower Manhattan, involved multiple architecture, planning, and design organizations that came together immediately following the attacks and provided a report and further input. The Municipal Art Society and a network of partners launched "Imagine New York" as a series of public "visioning" workshops, culminating during the week of April 11, 2002. People from neighborhoods and towns throughout the region came together to voice their opinions, concerns, and ideas for the future of the WTC site, the city as a whole, and their own communities in light of the new world order occasioned by the terrorists.

Downtown NYC was one of several interactive websites the public could access to put forth ideas as well. It was produced and managed by the nonprofit Project for Public Spaces in collaboration with the Civic Alliance to Rebuild Downtown New York—a coalition of more than seventy-five business, community, and environmental groups.

With input from Giuliani, Pataki formed the Lower Manhattan Development Corp. (LMDC) and wanted me to run it. But by then, let's just say it was pretty well known that Bloomberg's deputy mayor, Dan Doctoroff, and I didn't work or play too well together. They picked John Whitehead instead, a D-Day veteran and former chair of Goldman Sachs. He'd worked there with Neil Levin, the Port Authority executive director we lost in the Towers. I'd worked for Whitehead when he was Reagan's deputy secretary of state. There were some twenty board members on our committee representing various interests in Ground Zero.

With the LMDC, the Port Authority launched a comprehensive public outreach campaign called "Listening to the City." One of my top-notch guys, Glenn Guzi, a program director I hired right out of

Alfred University a couple of months before 9/11, wound up being the public face for that effort. He interacted with the community boards and the public. The idea was to distribute timely and valuable information through multiple advertising runs in daily, community, and foreign-language newspapers, as well as regular public service announcements. We widely distributed proposals and documents in several languages at locations throughout Lower Manhattan, as well as on the internet. And in the other direction, we sought maximum public input coinciding with each phase of our study of options for the WTC site, adjacent areas, and related transportation infrastructure[22]. The first public hearing took place in May 2002 at Pace University in Lower Manhattan.

We held several of these, including a high-tech modern town forum in July at the Javits Center. Five thousand people registered. (We could have fit more had I been allowed to expand the damn place five years earlier.) Along the way, we met with family members of victims; public officials; advisory councils; and community, civic, planning, and transportation organizations, including those representing New York/New Jersey PATH commuters. We believed the families of victims of the 9/11 tragedy, as well as public officials and transportation and community groups, must have a strong voice in the rebuilding of Lower Manhattan. They must be assured that a respectful memorial would be built to honor those who lost their lives, and that an improved transportation system would rise from the devastation caused by the September 11 attacks. We wanted to show the world that, by working together, we would revive and rebuild.[23]

"We had a clear directive from the governor," Guzi reminds me, "that, given the sensitivity of what had happened, we needed a very direct, specific, and diverse process for engaging the public. He was insistent we listen to the community in a meaningful way. 'Community' became something new to us, though, because it was a *world* community involved. So sessions went from 'Listening to the City' to 'Listening to the World.'"[24]

It was clear from the beginning that the public didn't want any version of Twin Towers II. People wanted something built, but something new and different. They wanted open space. Transportation. Street reintegration. Safety. Ease of access—how you get from one system like New Jersey's PATH to the city subways. And of course, we all agreed there needed to be a worthy memorial.

People say the public input process failed because it was all about debate. I say it was supposed to be all about debate. I don't want to say there's a downside to collaboration and public input, but it requires a lot of patience and goodwill to counter a lot of criticism along the way. A lot.

You have to include people. People's natural inclination is to participate in their communities and to get involved when problems arise. The so-called bystander effect, in which people in a crowd are less likely to step forward to help than if they are on their own, has largely been debunked.[25] The circumstances surrounding the horrific murder of Kitty Genovese in 1964, reported in the New York Times as "37 Who Saw Murder Didn't Call the Police," were inaccurately portrayed. Far fewer people witnessed anything; most heard just a distant scream. Two people did call the police, and one seventy-year-old woman ventured out and cradled the dying victim in her arms until police arrived. The New York Times admitted the erroneous reporting in 2016, in its obituary for Winston Moseley, Genovese's killer.[26] With few exceptions, people are inherently good and helpful.

In short, we live in a country where the government serves the people. So if you don't hear from the people, whom are you serving? Find them. Ask them. The public is the most important constituent in any engineering or building project. In general—and this is coming from a man who endured countless hours of public antagonism and recrimination—the public, collectively, is usually right.

The Public Runs Amok

Of course there should be a limit to public input. In 2016, the U.K.'s Natural Environment Research Council (NERC) offered the public the right to name its new polar research ship. When the polls closed, NERC realized it had a serious dilemma between its credibility and "the burden of public opinion." This was the problem: the votes overwhelmingly favored the name Boaty McBoatface.[27]

Apart from voting disasters by a desperate or ill-informed populace, I can't find any example of public participation's ruining a good project in America. In fact, the experience I've had with failures mainly occurred when we failed to properly involve locals in decision-making on public projects. The importance of a responsible media in spreading correct and unbiased information—not fake news—and a responsive government that listens to its citizens, particularly when an issue directly affects those citizens, is obvious.

Quarrelsome as it sometimes gets, public participation is valuable in gauging people's concerns, needs, interests, and willingness to try different solutions. Basically, it's better to know sooner rather than later how the public feels about a public building project, because the public will ultimately determine the project's success or failure. Better to have scrapped blueprints than buildings, parks, or infrastructure and institutions nobody uses.

Take High Line Park. From the 1930s to 1980, the 1.5-mile elevated industrial railroad linked what was once the Meatpacking District to the now developing Hudson Yards. Trains ran on it above Tenth Avenue, which had been nicknamed "Death Avenue" because so many people got flattened by said trains back in the day. Then it got shuttered by disuse with the growth of trucking. It became a thirty-foot-high, 126,000-square-foot concrete weed bed and garbage receptacle.

It was featured in Pat Benatar's 1983 music video for "Love Is a Battlefield." It was also a secret playground for locals, who'd put up Christmas trees and take amazed visitors to see it. It was like a scene from a zombie apocalypse film.

Some locals gathered supporters to fight Giuliani's plans to demolish the eyesore.[28] Bloomberg, who'd agreed that the West Side of Manhattan required vast redevelopment, gave developers $123 million of city money. Private donations amounted to a further $50 million, and the groups sought public input in a contest—glass-bottom pool, anyone? Roller coaster? At least no one suggested calling it Parkie McParkface.

The High Line linear park opened in stages from 2009 to 2014, along with phases of the overall Hudson Yards development—the largest private real estate development in the history of the U.S., and the biggest development in New York City since Rockefeller Center. The High Line quickly became one of the most visited attractions in the city, with nearly eight million visitors a year. "The High Line has fundamentally shifted the center of gravity in the city towards the west side and is expected to generate over $1 billion in tax revenue over the next 20 years."[29]

Can you call it a park, though? Or truly public? It's a cool repurposing of inner-city infrastructure, sure. But most of the typical programs and services that residents can expect from an urban park are verboten at the High Line. Surveys of local residents show they don't use it. They don't get it. It wasn't designed with them in mind. Instead, we've got a sort of pleasant "conveyer belt for tourists" through the trendy Meatpacking District or the soon-to-be-trendy Hudson Yards, through the condos and hotels along the way. There are a couple of hard-to-find access points in between those terminals for local residents to get on or get off. Our tax money was given less to a public park and more to a private-development showroom.[30]

There's a huge wealth divide in Chelsea, the neighborhood in which the High Line resides. The park looks beautiful, but does it really help the bottom line? My sense is that a true local public-interest campaign would have come up with a thousand other more pragmatic things to do with $183 million. We don't need another Celebration, Florida, or Dubai's World Islands. Aesthetics do matter—but only in service of practical utility.

The P3 Key

Having said that, the *idea* behind Friends of the High Line is just the kind of collaboration this country needs. It's a good example of a critical part of our re-engineering project—P3s. Legacy P3s consisted of government and private partners that together envisioned, paid for, engineered, built, operated, and maintained infrastructure projects such as ports, bridges, tunnels, highways, sewage systems, and water treatment control plants. A little imagination and ingenuity dramatically increased the reach of P3s across the world over the past few decades. What started as mainly physical construction projects has expanded into agriculture, health, education, civic and community development, telecommunications and tech, and, most recently, space exploration.

Probably the P3 that has attracted the most attention during this decade is, unfortunately, the prison system. Private corporations fenagled themselves into a badly broken system with the intention of massively cutting waste, increasing efficiency, and lowering operating costs. In a few cases this has worked. Mostly, it has been an embarrassing failure for the same reasons many government projects fail: greed, inefficiency, and graft.

Human (and therefore government) nature being what it is, of course these partnerships focus on the bottom line: profits. With few exceptions, such as the ones below, we can't expect pure philanthropy to solve our problems. There's no shame in the profit game. The best deals are always win-win-win (the third "win" is for the public). But now alliances between state entities and private corporate dollars have really shown us what they can do. Even social and environmental activism attracts strategic partners out of necessity.

Consider what the Mexican government's alliance with MTV and other telecom companies has done to combat human trafficking. Or look no further than the enormous philanthropic partnerships such as those undertaken by Bill and Melinda Gates, Warren Buffett, and Elon Musk. They've fostered partnerships with various agencies, institutions, and governments, with principled missions and tidy efficiency. The Gates Foundation alone has almost singlehandedly eradicated malaria. Millions of people are alive today who would have been lost to that deadly disease without that crucial partnership.[31] It took money, yes— Bill Gates–level money—but also the cooperation of multiple national and local governments across the globe—big governments, such as the U.K.'s, and smaller ones, such as Kenya's and Cambodia's.

Now if we can re-engineer a deadly disease affecting ninety-one countries[32] and bump it out of the top ten killers, don't you think we could repair some bridges and roads, and shore up some needed public services to improve neighborhoods? The point I'm trying to make is that we can't do it alone. We need private capital as well as government access and organization in addition to public (tax) dollars.

I agree with a 2013 report on the UN's Millennium Development Goals that our best hope for overcoming our social and infrastructural challenges is going to be the advancement of strategic unions—coalitions of the willing and daring—focused on joint missions and values. They call for "consensus-based decision making, shared accountability for outcomes and results, and synergistic interactions."[33] "Strategic" here means every partner comes to the table armed with a complementary bag of tricks—"unique resources to transform communities"—and a history of innovation and outside-the-box big thinking.

P3 collaborations are about more than infrastructure investment. Consider community policing. These partnerships between law enforcement agencies and the individuals and organizations they serve really work. Each side engages the other—not with guns and tear gas but with words and plans. Collaboratively, they develop solutions to problems and increase public trust in police. Trust is obviously lacking nowadays, regardless of the fact that most cops continue to be service-oriented, law-abiding, courageous servants of civilians. The police and residents of a community gather regularly to proactively and systematically study identified challenges unique to their communities. Then together they evaluate effective responses, thereby transforming policing organizations[34] from the ground up rather than the top down. At the same time, communities learn to work with—rather than against—the police. Body cameras and diversity training are just some of the many successful proposals that have come out of such meetings of the minds.

5

Invest and Monetize

Underlayment

The business community must be part of the solution to America's problems. The government alone will not cure our ills: the unemployed and homeless vets; the uninsured families; the disenfranchised poor; the hungry Americans. It's only by coordinating the resources of public and private power that every American's boots can get well and truly unstuck.

From my experience running the operations at Posillico, I can tell you with no uncertainty that work on foundations and infrastructure, generally below the level of our attention and daily lives, makes not only good business sense but engineering and political sense, too. It creates those opportunities for building new communities and businesses where none could have been before—like opening up the South Shore of Long Island just by laying sewer pipe.

The task of re-engineering needs to focus on all the underlying structures in order to build great things above, on the ground. We depend on water, air, roads, bridges, tunnels, and the dirt beneath our feet. Every one of those resources is essential, regardless of "identity politics," race, class, gender, sexual preference, or political persuasion. Whether we're teenage runaways or Wall Street CEOs, we need those basic structures in place before we can even begin to work on the higher goals, like better jobs, education, and harmony among races and genders. Everybody needs clean water (and wastewater that goes away and is processed quickly and safely); all people need good roadways to transport the food they eat and the goods they depend upon; everybody needs clean air to breathe and clean soil in which to grow crops. Those are our basic needs. We tend to take them for granted until bitches like Katrina, Sandy, or Osama bin Laden swipe them off the map for a time.

They're absolutely essential. The good news is we don't need a ton of new sewers, airports, bridges, tunnels, and roads. We just need massive maintenance done on much of our extant, aging infrastructure. The bad news is it's going to cost trillions. How do we pay for that?

Back during the Great Depression, FDR instituted the WPA, employing 8.5 million Americans over a span of eight years to bring electricity to the South; improve our roads, waterways, and utilities; distribute food and clothing; educate the masses; and construct and improve countless buildings, trails, tunnels, dams, bridges, and other structures. It was costly, but the government undertook the massive project out of sheer necessity, while the business community was struggling to recover. No one else could do what needed to be done, so the federal government stepped in.

Today, though, the economy hasn't been great for many middle-class households who grapple to pay their mortgages and pay off their student debts. But the business community—meaning big biz—has largely prospered. Many American corporations have profited in the ten years since the Wall Street crash, and the people at the top have brought home record paychecks and bonuses. So clearly, despite economic stagnation among the poor and middle class, it wouldn't be appropriate to think of this period as akin to the Great Depression; in fact, it might not even be stretching the term to call these past few years a recession proper. And since Trump's tenure, many aspects of the economy are booming.

So even though we have a lot of problems with our infrastructure, it doesn't follow that the government should step in to fix them. The WPA was an emergency measure at a time without any other options; America today has its fair share of emergencies, but the business community is thriving, so it simply doesn't need a government bailout. We cannot and should not rely on the government alone for this kind of progress. Instead, we have to look for P3 opportunities—for business leaders to chip in with their talent, money, and human resources.

Look at Chattanooga, Tennessee. Located in the Tornado Alley corridor, which makes it especially vulnerable to several types of natural disasters, Chattanooga was desperately in need of infrastructure development. Storms often led to massive power outages, which were costing the city upward of $100 million a year. After World War II, the population of Chattanooga dropped by 20 percent, owing to a loss of industrial manufacturing, and to suburban sprawl. When the Great Recession hit in 2008, city leaders had already been planning for years to upgrade their power and communications systems. With a combination of federal funds and city-issued AA+-rated revenue bond sales, the city launched a new electric grid that would save enough money to then offer a superfast universal broadband network, one of the most advanced in the country, to every home within a six-hundred-mile radius. They were able to attract new businesses because they were a public provider of electricity and internet services and were not forced to charge unreasonable rates for those services. By creating a "self-healing," smart-grid network, Chattanooga was able to reduce outages by 40 percent. Did it pay off? Ask the residents of neighboring Knoxville, who pay up to ten times more for their internet service. Ask the residents of Chattanooga, whose power after major storms comes back on after two seconds. Ask Chattanooga officials, who save about $1.4 million per storm.[1] When executed sensibly, these partnerships pay for themselves and then some.

Now let's head west. In 1994, the U.S. Forest Service developed a P3 with Recreation Resource Management (RRM), a family-run campground manager, to operate a busy public park in Sedona, Arizona. After more than fifteen years of operation, RRM did a case study to contrast the park they operated to a nearby state park that had about the same level of attendance and entry fees. They discovered that while the publicly run park cost the Forest Service almost a quarter-million dollars each year to run, the park operated by RRM returned nearly $45,000 each year in revenues.[2] That's almost half a million bucks a decade.

You could do the same with buildings. The James F. Oyster Bilingual Elementary School in Washington, D.C., was housed in an older, crumbling building that was in danger of closing. The public school district subsequently partnered with a local nonprofit group and a housing developer to use part of the school's land for a residential housing project, which provided funds to renovate the school building at no cost to taxpayers—not to mention provided housing for local homeless people. D.C. issued a thirty-five-year tax-exempt bond to accomplish the renovations, which the developer repays with annual payments.[3] Everyone makes money.

There are more examples across the capital. D.C. Water and Sewer Authority and its investors, Goldman Sachs and Calvert Foundation, have announced the nation's first Environmental Impact Bond, an innovative bond to fund the Clean Rivers Project. A few years ago, the partners began construction of green infrastructure to manage stormwater runoff and to improve the district's water quality. It's a $2.6 billion project that already includes a four-and-a-half-mile-long, twenty-six-foot-wide tunnel bored ten stories (one hundred feet) under Washington. It has attracted the attention of Elon Musk, which is always a good thing.

In New York, the Port Authority is overseeing construction of a world-class Performing Arts Center atop Liberty Park, next to the Greek Orthodox Church. Its stone edifice will be swathed in the same marble that graces the U.S. Supreme Court building and the Jefferson Memorial, sliced so thin it will be translucent. This could not have been possible without the Ronald O. Perelman organization, which has already contributed $75 million and committed an additional $25 million.[4]

Friends in high offices share with me all the time the success of their P3 projects in Texas, Virginia, and Colorado. It's an old idea—but it's the wave of the future.

Fill 'Er Up

Let's consider another possible solution to financing our punch list: raising the federal gasoline tax. Congress set it at 18.4 cents a gallon back in 1993, when the average price of gas was one dollar per gallon. It supports the Federal Highway Trust Fund (FHTF), which is responsible for financing road and bridge maintenance on a national level. It has stayed at that percentage to this day, despite the fund's depletion and rises in all other costs. Now that oil prices are lower, why not raise the federal gas tax to replenish the fund? This, along with leveraging some strategic P3s (private-sector financing), could begin to fund our infrastructure and social re-engineering project.

Experts estimate that a fifteen-cent increase per gallon (which wouldn't increase the percentage per gallon we pay in tax from its initial intent twenty-five years ago) would raise about $170 billion for the fund over the next decade.[5] It would prevent the current raiding of general funds and accounting "tricks," and also go some way toward curbing emissions and increasing the use of cleaner public transportation. Such a notion has bipartisan support.

A 2015 law will increase the FHTF by $20 billion over five years through traditional financing and a new slate of competitive grants—but that's not nearly enough. The law holds funding flat if you account for inflation. That's one reason why more than half our states have raised their fuel tax and/or sales tax—to fund transportation upkeep. I know this is an unpopular option among my Republican friends. I, too, believe the federal government should have as small a role as possible in the lives and businesses of the American people. But I know the other side of that bargain is necessarily a greater responsibility for businesses and state governments to ensure the continuation and improvement of our way of life.

Which gives us one of two choices.

The first is to consider raising taxes on corporations. Not by much—let's say 1 percent, just enough to cover the infrastructure those businesses use in transporting their goods, offsetting their impact on the environment and providing for their workers' basic needs. Just hear me out. According to the nonpartisan Congressional Budget Office (CBO) and the Joint Committee on Taxation, increasing all corporate income tax rates by 1 percentage point (say, from 35 to 36 percent for taxable income above $10 million) would raise revenues by $100 billion over the 2017–2026 period[6] from only the very richest companies. It would have no impact at all—except better services—for 99.99 percent of the public and the small-business community.

But the reflexive uproar this proposal would cause among my friends in the business world! "One percent? Are you kidding? You know how much that'll eat into our profits?" My response to this is simple. You're right. I get it. As a former executive at a profitable company, I don't sneeze at 1 percent of anything. But my company built roads and other infrastructure. So I'd also ask, how much would a closed road, a collapsed bridge, or a caved-in sewer system harm your profits? If your workers don't have clean drinking water where they live, or if your facilities have raw sewage backing up in the pipes, what are the odds your company will operate at maximum productivity?

The facts are clear: to survive and thrive, a corporation requires everything an individual needs, only multiplied manyfold. In fact, starting with the Fourteenth Amendment and proceeding through nearly a dozen Supreme Court decisions, corporations have enjoyed the rights and responsibilities of individuals. Why exactly shouldn't they pay their fair share? Well-to-do American private citizens pay upward of 40 percent taxes on their income—why should Facebook pay negative taxes?[7] More than half of Fortune 500 companies "have taken home a total of $27.3 billion over the past three years thanks to a tax break that allows corporations to treat executive stock awards like cash compensation—meaning the money can be written off like a business expense." Come on! And $76 billion in "accelerated depreciation" for Duke Energy, according to the General Accounting Office? Until a recent scandal involving Apple,[8] major companies held close to $2 trillion in profits parked in offshore havens, shielded from U.S. tax laws.[9] Who knows how much is still hidden?

Meantime, the roads, bridges, tunnels, sewers, and utility grids are crucial components of America's economy and thus individual *and* corporate requisites. If they cease to function properly, we all cease to prosper, and that isn't good news for business, either. Right now we have an opportunity to avert disaster; the business community can save itself billions of dollars in lost revenue by putting a few million dollars toward the construction and renovation of basic infrastructure today. Every individual pays taxes for this very purpose; isn't it only fair that businesses should have to do the same?

Maybe.

But there are significant downsides to raising corporate taxes to fund anything the government is in charge of. For every increase in taxes, there's another unnecessary project. Taxes supposedly go toward education, toward health care, toward housing, toward balancing government budgets. When the people—including corporations—subsidize education, the universities simply raise their tuition to cash in on all the free money that they're getting. When we subsidize health care, the insurance and pharmaceutical companies do the same. When we subsidize housing, the government builds housing "projects" that become blights on their neighborhoods, and none of their residents ever seem to work their way out of them. When the budget needs balancing, the government simply increases unnecessary expenditures to take as much as they can grab. So raising corporate taxes is not an ideal solution. The government can't be trusted to use the money for what it's ostensibly collected for.

All of that is another great reason for relying instead on individual state funds, matched by the government, for select projects the states' people want done and are willing to pay for through bonds and other investments. If corporate America wants to get a piece of that pie, they can do so as partners in a P3.

Inefficiency Experts

Sometimes investing means *divesting* your agency or organization of waste. Day one at the Port Authority—and super-unpopularly, I have to tell you—I started a years-long campaign to excise waste based on the model of Ronald Reagan. I started with a massive purge of excess personnel the Port Authority had accreted over the years. Despite a recent recession, we were paying a bloated staff of nine thousand, with many getting unnecessary overtime. In addition to a hiring freeze, I aimed to cut staff by at least 10 percent through attrition, buyouts and, finally, layoffs where absolutely necessary.

It wasn't the only trimming I did as part of that retrenchment. We also got out of the economic forecasting business, policy analysis, and managing a network of foreign trade offices, as well as curbed some controversial extravagances our executives had enjoyed, much to the public's dismay. I agreed with our new executive director, George J. Marlin, and his chief deputy, John J. Haley Jr., that our agency was militaristically organized (read: too heavily layered), and that there ought to be no sacred cows safe from the chopping block.[10] This sounds callous and inhumane, I know, but I sometimes wonder whether we could have afforded to trim more of our least valuable positions in service of that greater good. The idea was to free up capital to undertake the critical projects that were at the core of our mission.

By 1997, we'd saved about $230 million. Still, we had to abandon goals like the twenty-two-mile rail link between Manhattan and JFK airport—which I still, to this day, regret. It's embarrassing that New York is the only city of its size and importance without a proper express train from the airport to downtown.

There were other, bigger problems at the Port Authority. Our focus was always supposed to be on infrastructure, mostly transportation. Not real estate. Over the decades, though, somehow we got more and more involved in real estate holdings. That's not *necessarily* a bad thing insofar as the Port Authority is totally self-sustaining. No tax dollars support its $6 billion annual budget, which is more than many countries' entire budgets. The Port Authority is funded just by its own bonds and investment holdings, including real estate—and tolls and fares, of course. That's one reason it costs a staggering fifteen dollars cash to cross the George Washington Bridge, up from the original toll of fifty cents in

1931. Were it not for all the Port Authority's other investments, it would probably be five times that much.

But what's the *cost* of actually managing all those real estate holdings? Too much. We needed to privatize most of our holdings. We could make up any potential shortfalls by funding them in a partnership between New York and New Jersey, with help from the federal government. I'm not a fan of Barack Obama by any stretch of the imagination, but I supported his pledge to pay for half of the Gateway Project—basically under-the-Hudson infrastructure. The other half would have been split between the two states. Now President Trump's recent budget cuts that project off the rolls. Remember, he has promised $1 trillion-plus in infrastructure spending—but no one is entirely clear yet where and how that money will be spent—if ever. It derives mostly from unspecified spending cuts elsewhere. It's not real money yet.

In 2015, Governor Cuomo and Governor Christie set up an interstate panel to review the state of "cooperation and success" of the Port Authority. It was a shocking waste of taxpayer dollars. The commission determined retroactively that Port Authority infighting was responsible for some of the delays in rebuilding the WTC. Neither of those guys was there at the time. You can imagine how distressing it was to read that report fourteen years after the fact, knowing what I know about how complicated that job was, and how hard thousands of people worked to get it done despite impossible odds.

But there's one thing in that report with which I agree: the commission recommended the Port Authority divest itself of real estate development and get back to its core mission, which is transportation, especially improving our aging infrastructure. Yes. I was trying to do that the entire time I was there—and succeeding, slowly. Do I agree the Port Authority should be called "laughingly incompetent"?[11] That's taking things a little too far. It kills me.

But my three decades working for the government did nothing to lessen my sense that large-scale federal government operations are wasteful. As a general rule, governments breed inefficiency and waste in contrast to the forced efficacy of private corporations I've known or been a part of. So to get our tax dollars working right, we have to start savaging the overlapping and redundant processes, ineptitude, and blatant waste in government. It starts with cutting $400 Pentagon hammers (which saved the U.S. $106 billion the first year and helped Bill Clinton reduce the federal payroll to its lowest level since the Eisenhower administration)—and goes from there.

By unscrewing the tiny light bulb behind the big plastic display that covers almost the entire front of most soda machines—which serves no purpose but to make the can of Coke look more delicious—Texas saved about $200,000 a year in energy costs. (There are a lot of soda machines on state property!) Colorado used three different entities to deliver mail on the state office campus, including two government agencies and a private firm (proving that privatization alone isn't always the answer). You could literally stand outside the capitol and photograph three mail trucks following each other around from building to building. And West Virginia had never properly calibrated the salt-spreaders on its snowplows, so that whenever it snowed it was dumping far more salt on the highways than needed. Simply adjusting these devices saved the state about $3 million a year. None of these make a significant dent in structural deficits—but put together 100 small changes like that and, as the saying goes, pretty soon you're talking real money.[12]

Do you know how those states came up with those ideas? Do you think overblown and self-important government expenditure oversight review committees thought them up? Certainly not. They *asked* government employees across agencies for *their* ideas. Talk about a brilliant scheme for collaboration and buy-in. Those same employees—there are twenty-three million government employees in the U.S.—are usually the ones who have ideas about cutting fraud and abuse.

Contrast this to China, where the bosses would probably execute a government employee who was unproductive. I went to China in 2006, Pataki's last year as governor. The country was planning for the 2008 Beijing Olympics. The mayor, Wang Qishan, who was also the executive chair of the Beijing organizing committee for the Olympic games, took us on a tour. I asked him, "Mister Mayor, how are you planning to get the masses of people around to all the venues?"

He said, "Oh, subway lines."

I said, "Are the links completed yet?"

"We haven't started yet."

"You haven't started?" I said. "You have two years!"

"Plenty of time. *Plenty.*"

Pataki and I talked about it on the plane on the way back to New York; he just kept shaking his head. It would take us fifty years to build a tenth as much track—a twentieth!

Sure enough, though, it was all open by 2008. Even today, crews are still burrowing, tunneling, and banging up new stations and laying new lines 24/7, 365 days a year in Beijing. Between 2007 and 2014, the megacity added 235 miles of new routes—a length greater than the entire New York City subway system.[13] They're twice as long as ours now, and only Shanghai's subway is longer.

How do they do this? Simple. Their workers are cheaper, there's almost literally an endless supply of them—and they have virtually no regulations. Those they do have, they just break. The environmental ones, especially. I was recently in Shanghai. We were going to go to Beijing, but we had to cancel the trip. People there couldn't breathe. They advertise the malls there as sanctuaries from the smog-choked air outside. It's an unsustainable compromise that the Chinese had better get a handle on soon. Then again, maybe they think they can afford to lose a few hundred million people to lung cancer—there would still be "plenty."

Life Partners

In far too many cases, public sector jobs become a real drain on the economy, and the only solution to that is to replace public jobs with private ones wherever possible. There isn't an American older than six who doesn't know that UPS and FedEx are ten times better at delivering packages than the federal government.

"Well, wait a minute," my business buddies might say. "First you say you want to raise taxes, then you say you want smaller government. Which is it?" My answer to that has got to be *both*. As far as I can figure it, from all my years of experience as a businessman and as a public servant, there are only two ways we can rebuild our infrastructure to meet our basic needs and then move on to tackle our more intractable sociopolitical challenges: either we create an airtight funding system within the state and federal governments whereby we must use certain taxes for the purposes for which they're raised; or the corporations need to organize these building projects themselves. The middle ground is P3s, which have closed significant gaps in Australia, Africa, Asia, and across Europe.

And for all the problems that government presents, I tend to believe that the former option is the more sensible solution. Why? Because most corporations don't know anything about specific infrastructure. They use the roads, the sewers, the power grid, the same as all the rest of us, but the vast majority of businesses have no greater understanding of the workings behind individual resources than the average orthodontist on the street. In general, what does AT&T know about water treatment? What does News Corp. know about airports? What does Union Pacific or CFX know about internet bandwidth? My guess is, probably not too much, and if these companies needed to helm a project constructing or rebuilding major American infrastructure, they'd sooner pay someone else to do it than create a deep sewers division of, say, the Adobe Systems team.

That, in effect, is what taxation is meant to be. You pay a public service agency to deal with all the tasks you require to maintain your way of life, but about which you don't know much and probably wouldn't want to deal with yourself. Getting all the corporations to act in tandem would be a nightmare. But getting one or two onboard to partner on a specific project in which they see purpose, proficiency, and profit—that kind of strategic alliance is what P3s are all about. Healthcare. Renewable power. Urban and transnational transportation. Ecofriendly water delivery and waste consignment. Environmentally sustainable parks and recreation. All of it is more financially feasible with private partners.

We've gotten used to T-Mobile Arena, AT&T Park, Busch Stadium, the Mercedes-Benz Superdome. We could live with General Motors' I-25 and the Tesla Midtown Tunnel (both better than "Hershey's Highway"). New, relaxed rules for corporate donor recognition could soon expand branding to national parks and historic sights. Can we get our heads around "Disney's Grand Canyon"—or is there a distinctly un-American ring about that? Enough to prevent billions of dollars in upkeep in exchange for more eyeballs on private brands? The cash-strapped National Park Foundation has a $12 billion backlog of repairs and projects with just a $2.85 billion annual budget.[14] Trump's Ford's Theater, anyone? We've already got Trump's White House.

Look, this is just an extension of the direction in which we're already going: all ads, all day long on those high-resolution, high-impact LCD/LED/LPD video walls like the ones in the Transportation Hub concourses. This old-timer feels like I'm in a sci-fi movie that's set in space, passing through a tube into Jupiter Customs and Immigration.

But monetizing in this way can help innovative engineers keep costs of new developments extremely low. For example, Hyperloop Transportation Technologies in California is proposing that within a few years we'll be able to travel overground on magnetic levitation capsules through low-vacuum tubes at the speed of sound—without having to buy a ticket. Seriously. They're already building these things in the Great Lakes region, France, South Korea, China, the UAE, and Slovakia. But why might the system be free for commuters? One way is that operators will monetize the interior walls of the ninety-eight-foot capsules (twice the length of a subway car). L.A. to San Francisco in half an hour—sponsored by a new Netflix cop show pilot.

The monetization possibilities for Hyperloop's stations are unlimited. Would we turn down a thirty-minute commute from New York to D.C. if the transformative infrastructure bore the name of Company X? We take named airlines now. We've accepted iconic structures, like the Chrysler Building, named after private citizens.

If we can avoid obvious embarrassments like "The Statue of Liberty—Sponsored by Summer's Eve®," or "Trojan® Brand Condoms

Washington Monument," or "Canyonlands National Park—Brought to You by Vaseline®," we might seriously consider extending such branding partnerships to cultural and historical sites.

The only other option is to totally reform our tax system. We'd have to disband the IRS. Good riddance. We'd need to form and deputize some government body to coordinate a proportional tax on all corporations. If there were a legal means to ensure that monies and project costs coincide, we could have a partial solution that would benefit all parties. Both the public and the business community would win by getting improved infrastructure and averting costly disasters down the road. We'd also create more jobs and put masses of people to work, thereby creating a greater demand for consumer goods, which, of course, is good for the economy.

It would probably be wise for people to demand the creation of the legal means for enforcing some stricter tax-allocation rules, even if it's just for income tax and sales tax paid by the public. If we want a way of forcing tax money to be used only for certain preconditioned purposes, we have the Constitutional capacity and some precedent for petitioning it. It isn't an easy feat, but that's exactly what democracy demands—with the liberty to choose comes the responsibility to make the right choices.

Let government, private sector businesses, and the public each do what it does best, and each contribute its fair share. When all parties, including the public, stand to benefit, let us partner to optimize our resources. Together we can re-engineer a future for our country based on a shared and smart investment.

Sooner or Later, You're Talking About Real Money

The janitoring business was pleasantly profitable during the war years. My father was making about one hundred dollars a week, which was a lot of money at the time and not bad for a temp job during the construction hiatus. He had turned a potential negative—the shutdown of his industry—into a positive. He adapted his skills, got ingenious, and remained industrious. This was typical of the immigrants in America.

They didn't bitch about the circumstances or the economy; they just found something else to do. They did what they had to do. What other choice did they have?

There are certainly immigrants today who work just as hard and get just as entrepreneurial and inventive. Many support their families here or back home by working several menial jobs. I've met cabbies who were dental surgeons in their home country, maids who were mortgage brokers. But I'm not sure that's the norm anymore. Laziness is not a uniquely American problem.

My father, on the other hand, earned enough to buy our brownstone on Seventh Street, the place on earth I felt safest and happiest. That would have been a good enough investment on its own, but he soon discovered the power of equity—money out of thin air. He borrowed from his local bank to get a down payment on an investment property on Third Street. It's one of the oldest and most effective tactics you can use to grow your own wealth. It's not risk-free, but almost. He rented out those apartments and paid the remaining mortgage with the rent money. He had enough left over to pay down the principal and gain equity in the first investment property. And with *that* money, he put a down payment on a second investment property, and a third, and so on. I've watched people, including immigrants in Harlem who began in near destitution, employ the same technique to grow enormously wealthy. I know I sound like an infomercial, but it's true.

A roof over your head tonight is urgent. But your family's enduring financial health is important. It's either home sweet home or home sweet investment. Over the past two centuries, U.S. real estate has been the best intermediate- and long-term investment you can make. As long as your version of the American Dream fits in your budget, you do your due diligence, and you prepare for contingencies so you can hold on for the 1,776-foot long term, it's the closest thing you can get to a guaranteed return. Meanwhile, you provide your family a home they will cherish, as I do the first Seventh Street brownstone—and the first house I bought with my wife, Prudence, when I was twenty-four in 1960. That was in Deer Park near where my parents kept a summer home. My first son, Larry, was born there in 1961. We sold that house five years later to build our next home in Dix Hills, in a neighborhood I fell in love with while developing a golf course for Posillico. My daughter Carla was born while we lived there.

Yes, there are a few specialists who make a living and a few who get rich "flipping" properties. I'm not talking to those people. I'm talking to the person stuck on the hamster wheel of too much month at the end of the money. Do whatever you have to do to invest your money. Seems counterintuitive, right? People struggling financially don't have the means to invest in the future because what about keeping the kids fed and clothed today? That's the exact kind of mentality that keeps that hamster wheel spinning for so many Americans. If you keep doing the same thing over and over expecting different results, that makes you insane, not responsible.

Or you could just rely on the handout, the instant gratification of a monthly check from the government *because* you're poor and always will be.

We encourage that kind of attitude in certain communities. Guaranteed government benefits for single mothers with five or six children from different fathers is a good example. Sometimes I think those mothers do it because the more children they have, the more they can collect. They openly admit it. I'm not picking on a particular race, but it's a reality that there are a disproportionate number of broken families of certain ethnicities. This is a sensitive area, I understand, but maybe if they didn't get that compensation, they would think a little more about their choices.

I assume you can tell by now that I have zero tolerance for that kind of idleness and irresponsibility. I put on the important list getting folks off welfare through vigorously run and monitored back-to-work programs, many of which have gotten gutted in some states over the past few years. We need to invest in our compatriots' futures. One of us drowning means someone on solid ground needs to do something about it—not throw a lifeline but offer swimming lessons.

Millionaire, Mogul, Mensch

Consider my friend Max Fisher as a model for smart investments. He's the one from the Bush-Clinton-Perot debate. He was born to Russian immigrant parents in a middle-class neighborhood of Pittsburgh. A year before he died at age 96 in 2005, Fisher had amassed a net worth of $775 million. He was the oldest member of the Forbes 400.

He started working for his father's oil reclamation business for fifteen bucks a week. Fisher made his fortune stock-prospecting and in Detroit real estate. Then he gave millions upon millions to worthy causes, ranging from his alma mater, Ohio State, to his beloved State of Israel. It was for that reason he rose to prominence as the elder statesman of North American Jewish activism, and caught the attention of world leaders. For decades, he served as a trusted adviser to U.S. presidents and Israeli prime ministers. His major purpose in life was to forge stronger ties between America and Israel. He dedicated his life and the bulk of his money to that cause.

See, financial wealth is only one part of the equation. There are many ways to be wealthy. Perhaps the best—and the one that's least likely to trample on others—is to be rich with a purpose like Fisher. Once you have that, do everything in your power to continuously reinvest in that purpose. Don't let trivial matters get you off track. If you can turn that purpose into service, you'll feel like the richest person on the planet. I followed Max's example and endowed a chair at Stony Brook University.

Along the way, invest in some presentable clothes. Pay attention to appearances and impressions the way Fisher did. He was a former linebacker, but he always looked classy, like an elderly James Bond. The way you look and dress might sound trivial in the grander context of your mission and values, but it's not. It's the outward representation of who you are inside. My boss at the *Eagle*, Mr. Cavagna, wasn't a big, handsome guy, but he was always well-dressed. You don't have to be wealthy to be well-groomed, neat, clean, and, for work and special occasions, even natty. I've always taken to heart the old advice that you should dress for the job you want, not the job you have. You just feel better when you know you look good.

Of course the substance underneath the shell is most important. But can you imagine anyone would be impressed with the new WTC if it were hideously clad in popcorn stucco? The reason we polished the mailbox and swept the streets in front of the house on Seventh Street when we were kids is because my father had fierce pride of ownership. For the same reason, cleaning up—literally—the streets of New York since the 1970s has proven a positive first step in improving neighborhoods, lowering crime, decreasing vandalism, and increasing people's morale. Even the much-maligned stick-a-flower-pot-decal-over-a-busted-Bronx-windowpane plan of the early '80s worked as a cost-effective measure. The mayor used a federal grant of a mere $300,000 to improve only the derelict buildings' appearance—as opposed to the tens of millions it would later take to actually rehab those areas. Perception is reality, as a housing official said at the time.[15]

Are You Talkin' to Me?

Let's put some of these pieces together: listening, leading, collaborating, partnering, and investing. That's exactly what we were trying to do with the founding of the Tribeca Film Festival. The year after 9/11, Robert De Niro and partners Jane Rosenthal and Craig Hatkoff came to me with a proposal they'd been trying to get off the ground to revitalize De Niro's beloved "**Tri**angle **Be**low **Ca**nal" Street. I thought a film festival was a fantastic idea—and the timing couldn't have been better. We all found ourselves in desperate need of getting things moving again downtown. I'd always been a huge fan of the movies. *Gone with the Wind*, of course. And my favorite, John Ford's *The Quiet Man*. George Pataki and I and our families recently stayed at a friend's castle in Ireland, where we walked the streets and strands of the tiny village of Cong on the Mayo-Galway border, where John Wayne romanced Maureen O'Hara on his character's ancestral lands—magical.

Of course I'd always admired De Niro—not that I necessarily agree with all his politics—ever since *Mean Streets*. We became friends, and I was able to secure him $200,000 from the state through the ESDC. He helped a lot, too, raising a ton of money in addition to his own. The first year attracted 150,000 people, and there were twice as many the second year. Since then the film festival has expanded in size and scope to become one of the biggest and most important in the world. I used to go with Bob every year to the opening, say a few words.

If you want to go now, get ready to fork out $25,000 a table. Now the festival honors not only films but TV, music, live performances, video gaming, and even virtual reality programming. All that started because Bob had a vision for improving his neighborhood, a skill set to make it happen, and the ability to schmooze the right collaborators—me included.

This is going to sound like I'm tooting my own horn, but I'm trying to make a point. As an ancillary benefit of my association with De Niro and his comrades, I got to appear in five feature-length films, starting with *The Devil's Advocate* in 1997 and finishing with *World Trade Center* in 2006, in both cases playing myself. I also played FBI agent John Kravec in *Witness to the Mob* (1998), a golf announcer in *Serendipity* (2001), and a maître d' in De Niro's *Analyze That* (2002). Not that any of that makes me any cooler in my grandkids' eyes.

The Unicorn of the United Front

How do you get the kind of buy-in Bob and I got for the Tribeca Film Festival? First, you need a common mission that everyone is supportive of and interested in. The surest way to destroy any project is to have the wrong people on board, people who are uncommitted, inefficient, and don't care whether or not the thing succeeds. The greatest leader in the world can't bring a project to fruition without the strength of teamwork behind him or her. That's as true of a building project or business/organization as it is of a country.

It would seem that diversity automatically leads to dissent. That's often true. But how do you ensure productive dissent doesn't turn into backbiting, big egos, people at cross-purposes, laziness, wasted talent and time, active thwarting, sabotage, and personal agendas at the forefront? In short, you don't. You can attempt to minimize these things, but you'll never achieve the perfect team.

I can think of three examples with which we came close to the ideal, though—models of good collaboration from a large, diverse team that resulted in obvious success.

Of course the WTC rebuild is first on my list. But that took a nightmare to precipitate action. You remember those first days after the attack? Everyone came together. Every man and woman I ever met on that Pile shared the same mission: first to rescue, then to rebuild.

But why must it take a terrible disaster to trigger us coming together for the common good? It doesn't always happen that way.

Consider my second example, Central Park. In 1857, Frederick Law Olmsted and Calvert Vaux developed the Greensward Plan, which was selected and approved by the Central Park Commission appointed by New York state. The execution of the plan depended on a few key individuals: architect Jacob Wrey Mould, master gardener Ignatz Anton Pilát, engineer George E. Waring Jr., politician Andrew Haswell Green, and designers Olmsted and Vaux.

They cleared the land of about 1,600 squatters and their livestock (an action the team deemed necessary for the greater good of the city—there's that eminent domain again). Then they purchased an additional sixty-five acres at the north end of the park and started work in earnest. Using a combination of up-to-date steam-powered equipment and custom-designed, wheeled tree-moving machines—as well as twenty thousand unskilled laborers with just shovels, carts, and 166 tons of dynamite—the team built the park. They worked from 1860 (during an economic depression and the Civil War) to 1873. Haulers dragged more than ten million cartloads of soil and rocks from the park—I wonder what happened to all that dirt—then lugged in 18,500 cubic yards of topsoil from New Jersey. The gardening team transplanted four million trees, shrubs, and plants representing about 1,500 species. Blasters used more gunpowder to clear the area than the Confederate and Union armies combined at the Battle of Gettysburg.[16]

The project was not without its tensions—particularly between the designer Olmsted and the politician Green. It was not unlike the architect wars of the WTC rebuild, with each man's vision garnering an army of champions and an equal number of decriers. Despite the tensions over leadership, the park itself simply could not have been built without the teamwork and shared vision of these men, and the tens of thousands whom they employed.

My third example is probably going to come as a surprise to some. Again, its antecedent was a disaster, this time a natural one. New Orleans. Within days of Hurricane Katrina, the Army Corps of Engineers repaired the fifty-five levee breaches. Within one week, the Port of New Orleans was able to receive and service relief ships, and crews restored power to select buildings in the central business district. Commercial shipments resumed two weeks later.[17] Water and sewage services gradually came back in various sections of the city, with the last section, part of the Lower Ninth Ward, fully repaired more than one year later (October 9, 2006) after countless, tireless hours served by crews in sludge filled with tree branches, swift-moving wastewater replete with "turtles," and clouds of deadly methane.

About 80 percent of the city, which lies below sea level, was underwater as deep as twenty feet. The storm left a $108 billion trail of destruction. In the aftermath of Katrina, the Department of Homeland Security under Director Michael Chertoff managed the deployment of fifteen thousand active-duty members of the military, forty-three thousand National Guard members, four thousand Coast Guard personnel, and seven thousand FEMA responders.[18] While reconstruction has been uneven, with the hardest-hit, low-income areas remaining in ruins the longest, the city has been rebuilding through active community participation and P3s.

Frederic Schwartz, the architect selected to replan one-third of the city for 40 percent of its population, noted, "Disaster offers a unique opportunity to rethink the planning and politics of our metro-regional areas—it is a chance to redefine our cities and to reassert values of environmental care and social justice, of community building and especially of helping the poor with programs for quality, affordable, and sustainable housing."[19] This included making "every effort to involve the residents and the community in the planning effort" and to "preserve the unique architectural heritage of New Orleans," with the rebuilding based on the model of surrounding neighborhoods wherever possible.[20]

We managed something similar in the rebuilding of Ground Zero. One of the most consistent and vocal responses we got from the public in the Ground Zero rebuild was to restore what's called "street grid integration." For forty years, the discontinuity of the modernist "superblock" of the original WTC complex had really pissed citizens off. I remember that part of Lower Manhattan as it had been before the late '60s, before the web of local streets you could walk was razed in favor of dead ends and impenetrable walls bounded by widely spaced, higher-speed, arterial roads meant exclusively for cars.

We kids would come out on the IRT from Brooklyn to shop for stickball bats and gloves at Modell's on Cortlandt Street, while my mother shopped for stockings or whatever. Meanwhile, my father and brother rummaged the stalls of Radio Row, dominated by the max-packed, labyrinthine Heins and Bolet store. They were hunting for some complicated oscillator, coil driver, or tube to get our Zenith Stratosphere—our home's electronic hearth—up and running again in time for *The Abbott and Costello Show*.

The Port Authority leveled some 144 buildings in that Lower West Side neighborhood in the name of eminent domain to make way for the brashest new development of the '70s—the World Trade Center.

Anyway, it was loud and clear at our public hearings that people missed the former street grid. People wanted us to reintegrate the new WTC into a proper "neighborhood" redevelopment versus some sixteen-acre superblock with more fortified walls. There's a pretty huge distance between, say, Broadway and West Street—all kinds of room to restore some of the old streets for ease of access and more of a sense of community.

That naturally raised terror alarms, but we did our best to think about integrating all that into our overall architectural plan. For instance, the visible ribs in the long concourse of the subgrade entrance from the Transportation Hub into 1 WTC are actually holding up the restoration of Fulton Street above. We also restored Greenwich Street, a north-south lane between the Transportation Hub and the site. It feels much more neighborhoody to me now.

Just as at Ground Zero, in New Orleans thousands of individuals and businesses have been involved in reconstruction, funded by dozens of charity and nonprofit organizations, such as the Rockefeller Foundation, the Greater New Orleans Foundation, Habitat for Humanity, and the Bush-Clinton Fund. The city is now also safer from storms, owing to a $14.5 billion hurricane and flood protection system.[21] More than thirteen thousand blighted properties have been demolished or brought up to code, and $1.63 billion has been invested in new roads, parks, playgrounds, and community centers.[22]

None of that, by the way, is thanks to federal prisoner number 32751-034, Ray Nagin, former New Orleans mayor and frequent critic of mine. When he said, in August 2006, "You guys in New York can't get a hole in the ground fixed, and it's five years later," I wanted to punch him in the mayoral schnoz. My colleague Kevin Rampe, chairman of the LMDC, put it more mildly. He reminded the future prisoner that tremendous progress had been made in lower Manhattan. For God's sake, there were 160 major buildings cloaked in debris—there was dust on the Empire State Building, three miles away! Now the Freedom Tower, the Transportation Hub, and a memorial to the nearly three thousand attack victims were under construction. He said *we* knew better than anyone how difficult it is to rebuild a city after destruction.[23]

I was not as patient. The next day, mad as hell, I brought the whole national media—heads of the boards of all the major papers—to Ground Zero to show them all our progress. "See there," I said, "seventy feet down full of infrastructure. Trains running. Highway running. River kept out. We have to get everything down here right as we go or this gigantic puzzle won't fit together. There's emotional urgency, sure, but we have to get all this right, you understand?" They all nodded. They understood.

Was the press the next day good? Take a guess.

More P3 Magic

The ideal public works project is one initiated *by* the public and *for* the public: say, a park or a playground, or the rebuilding of roads, bridges, or other infrastructure. The public is never quiet about the things it wants or the services it feels are in need of improvement; just ask the staffers of America's lawmakers. According to a recent study, lawmakers are frequently bombarded with public grievances, though "the best way to influence a member of Congress is to visit them in person; and calling their office is less effective than you might think. Instead of picking up the phone, concerned citizens should write a letter to their representative, or a letter to the editor. Email is effective, but *form* emails aren't."[24] (Emphasis mine.)

Even better is when nonprofits and government agencies reach out to the public as we did in our Listening to the Planet and Neighboring Galaxies campaign.

Public participation has completely altered the landscape in Santa Ana, California, which provides a perfect model for how citizens, nonprofit organizations, and government can work together to solve problems that affect everyone. In Santa Ana, an impoverished neighborhood in the midst of extremely wealthy Orange County, California, there were no public parks, and children were forced to play in parking lots after school. Doctor America Bracho, executive director of Latino Health Access, helped organize neighborhood residents to collectively decide on a solution, and together they obtained a $3.5 million state grant to create the half-acre Green Heart Families Park and Community Center.[25] In addition to that park, Bracho has helped the community organize health education and access—hundreds of volunteers who visit and assist diabetics, offer cooking and exercise classes, and provide translation and other necessary services for fellow residents.

Just as with a business venture or nationwide re-engineering, having a team dedicated to the project and invested in the results makes all the difference in the world. Members of the public are not something to be feared or skirted around when re-engineering the country is the goal. They're the ones who know every pothole in the road, know every solution that didn't work, and more likely than not, have a pretty good idea of what ought to be done. Collaborating with the public is always a smart move where public works are concerned. The sooner you listen to what the public is saying, the less heartache you'll have down that potholed road.

Stressing the importance of collaboration and adaptation may seem counterintuitive, but every great leader I've ever met or read about has understood teamwork to be an invaluable contribution. The common idea of great leadership tends to overemphasize charisma, that take-charge attitude, and gut instinct, but while those traits can certainly come in handy, they don't mean a thing if you're pushing for the wrong solutions—or pushing against popular opinion. Take the time and have the humility to listen to the right people. Nobody knows everything, no matter how much experience and insight you think you have. Knowing what you don't know and how to find it out is far more useful to everyone than your ability to be beguilingly bossy but wrong. Find the best people you can for your team and have the intelligence to listen to them. Encourage them to voice their concerns, give them the credit they deserve, and unite them for a common goal. As Andrew Carnegie said, "Teamwork is the ability to work together toward a common vision. The ability to direct individual accomplishments toward organizational objectives. It is the fuel that allows common people to attain uncommon results."[26]

I Just Can't Work with This Kind of Guy

On March 15, 2006, I held a pissed-off press conference at the Port Authority's temporary headquarters and told the world that developer Larry Silverstein was a "very greedy man." The following day, Pataki told the press that Silverstein, a former ally, was a "traitor" who'd betrayed the public's trust.[27] He'd put his own financial interests ahead of New

Yorkers' and all Americans', and tried to profit from the worst tragedy perpetrated on American soil. I don't like publicly shaming people; it's not in my nature, and my mother would not approve. But even Silverstein's confederate and my sometime nemesis, Democratic New York Senator Chuck Schumer, got sick of Silverstein's avarice. Here's what led up to my statement.

Between our two states, New York and New Jersey have had nine different governors since 9/11. New York's four different governors appointed five executive directors to the Port Authority, and the five New Jersey governors have appointed four separate chairpersons. In the end, most of the major decisions were made by Pataki and me while we were still in office. The only major thing not completed when Pataki left at the end of 2006 was 1 WTC—obviously a big part of his legacy. Having seen firsthand what I had done for 42nd Street, Pataki told me he didn't trust anyone but me to get it done. So in 2006, before he left office, he put me up for an extended term as a Port Authority commissioner. The state senate approved and confirmed me to serve an additional six years. I was no longer the vice chair because the new governor, Democrat Eliot Spitzer, got to pick his own—when he wasn't choosing an escort from an expensive agency under the watchful eye of the Feds. But Pataki's exiting orders to me as a commissioner were to get 1 WTC built.

This required me to make a deal with Silverstein. Remember, the Port Authority owned the land, as well as the PATH train that linked New York and New Jersey, and we'd sold the WTC site to Silverstein only three months before September 11. For the record, I supported privatization 100 percent, so we could work on getting back to our core mission. Of course we had no idea that just a few months later, that privatization would cause one of the worst imbroglios in real estate history.

I was heading the Trade Center Development Committee for the Port Authority. We were to decide what to do with the WTC site, what to build there, the whole shebang. Nobody except Larry Silverstein thought Larry Silverstein was capable of building what was then called the Freedom Tower without significant help from us.

So, near the end of his tenure in 2006, Pataki and I, along with Tony Coscia, the Port Authority chair from New Jersey—a real standup guy—met at our temporary headquarters. Larry Silverstein was due to arrive at 1 p.m. He came at 2:30. We wanted to know from him what he'd sell the Tower 1 project to us for. The governor and I had worked out a number in advance—$250 million to $275 million. We gave Silverstein the number. He'd get to develop buildings 2, 3, 4, and another proposed building—the most commercially desirable properties on the whole site—and we'd build the Freedom Tower—Tower 1, which would eventually become 1 WTC. In turn, Silverstein would turn over $1 billion from the insurance policy he'd been paid. He said, "Well, we've gotta work on this all day. See you later tonight."

We gave him till 6 p.m.

So we waited. And waited. We ordered Chinese. Finally, Silverstein and his team came back at twenty minutes to midnight. "A billion," he said.

"Hold on a sec," I said, and conferred with our executive director, Lew Eisenberg, in another room. A billion dollars! This slimy Silverstein was trying to shake us down. He wanted even more of the insurance payout to cover his construction costs—but wasn't willing to cough up anything for our costs. He was also demanding lower rent—he was only a tenant; we owned the site. Not only that, but he also demanded we double the $350 million rent reduction we had offered on his $125 million-a-year ninety-nine-year lease. "Show him out, please," I told one of the other commissioners. Or maybe I said, "Shove him out."

Then I held that press conference the next day. It was a poker game with the highest stakes—and Silverstein was holding jack shit in his hand. I called his bluff. I publicly announced that unless the developer returned with a reasonable offer, the Port Authority would require that he go forward with his contract and begin building the $2.3 billion Freedom Tower starting the next month. "For Silverstein, that [was] the building that cost the most and [would have been] the most difficult to lease—and the state would withhold hundreds of millions of dollars in Liberty Bond funding"[28] if he didn't play ball with us.

I wanted him to either play fair or get out of the way. He held his own press conference saying we'd angrily abandoned the talks—he had been prepared to hash it out all night, no matter how much coffee or how many lawyers it took, but we kicked him out for no reason. We had a reason.

Silverstein blinked. Bullies always do. The Port Authority spent less than $400 million on the deal, and he ceded the rights to us to build Tower 1,[29] the most significant structure on the site.

I was used to seeing such naked unscrupulousness. After 9/11, we received money from the federal government, several billion dollars, and it came through my agency, the ESDC. This was beyond the $7 billion September 11th Victim Compensation Fund, which awarded on average $1.8 million per family to 97 percent of the victims' families in exchange for a promise not to sue. Of course we had a whole process for divvying up that loot, a typically bureaucratic nightmare of paperwork and triple oversight—think DMV, then quadruple the pain. We determined need based on brunt. All kinds of complex and interrelated factors went into this calculus, starting with distance from Ground Zero, a good gauge of impact. We used a series of four concentric circles to delineate zones of impact from Ground Zero up to Canal Street.

We found that the majority of entities were honest—individuals who lost their apartments or small businesses, and large corporations like that of my Bridgehampton neighbor, Howard Lutnick. He ran Cantor Fitzgerald, which lost 658 of its employees, including Lutnick's brother, and suffered lesser, but significant, financial challenges in the wake of the attacks.

But some companies, such as Con Edison—which, granted, did take hits—pushed to get a disproportionate share of the emergency funds.[30] Sometimes tragedy brings out the worst in human and corporate nature.

The Brainpower Piggy Bank

That Andrew Carnegie had some melon on him. He reminded us that investments aren't necessarily financial. That's cold comfort to me now in light of the hugest financial mistake I ever made.

Remember the pennant-winning Dodgers baseball with all the authentic signatures I got when I was thirteen? One day, the Seventh Street gang couldn't find another ball to play with. I went to my room and we used that ball all afterno0n—wore down all the signatures. I recently googled what it might be worth today. You don't want to know. We had fun, though.

It's not all about money. I mean it. Education is more often than not a good investment in the future. While a college degree is by no means a ticket to getting a job anymore—and while college is definitely not for every student—as a general rule, investing in education pays off in the end. This could mean vocational training, union training, on-the-job experience, and/or traditional higher education—just don't fall under the liberal bulldozer. You also could take advantage of the literally unprecedented—and free or virtually free—opportunities for online learning.

The idea is to invest in yourself in order to increase your chances of success. This will raise up your family and ultimately, the nation. All boats rise when anyone contributes even a drop into the ocean. Get yourself in a position to teach others how to swim, and occasionally, selectively, when necessary, throw out a worthy lifeline. For all my fear about the future of Millennials, many are poised to retire in their forties because of good educational investments, wise real estate transactions, old-fashioned scrimping, and the philosophy "Yes, live in the now, but plan for the future."[31]

I'm In

The penultimate kind of investment I want to talk about here is a national one. The giants on whose shoulders we've stood invested their blood in our freedom. The least we can do is reinvest ourselves in the American values they fought and died for. We get paid the dividends every day. But what do we invest?

I've talked about "buy-in" before in a different context. But when a group, under good leadership, shares its core values, signs on to the mission, agrees expressly and implicitly with the benchmarks and goals, what happens? What would happen if, as an organization, a family, a business, a government, a nation, we all had a say in formulating the clear expression of those values? If there were a mission statement that everyone bought into? That's exactly what the Constitution and its amendments are. And what our Declaration of Independence is.

Even after the bickering, infighting, and public criticism began, the WTC team continued to share essential values—we were invested in the same core mission. We would rebuild the site. We would make it better, grander, safer. We would keep half of it—including the footprints of the original towers—as consecrated ground to honor the dead, even while building the other eight acres into world-class commercial space to exalt our resistance, our spirit, our power.

A huge benefit of a values-based mission statement is that it goes quite a way toward keeping people less invested in their own egos. When two or more stakeholders disagree, even vehemently, they can both go back to the mission statement and make a decision based not on their personal tastes and preferences, but on the already articulated, clear, values-based mission statement. "We all agreed and should stand by our commitment to keep half the site as a memorial. So is what you're suggesting in line with that mission or not?"

Of course people of good faith can still disagree, even in the presence of the mission statement. Our Constitution is one of the clearest, most ingenious documents ever conceived by the hand of man, but it still requires some level of interpretation for novel and evolving circumstances. Our Supreme Court is meant to stay politically impartial, and all the decisions of the nine justices are supposed to be made in a kind of vacuum of pure Constitutional law. Does it always work that way? No. Should it? Ideally.

As a country, we need to get back to investing in the primary values on which America was founded, which carried us through both dark and bright days. Our founding fathers struggled tirelessly for nationhood, battled the most powerful forces on the planet. It was a clash of values, wasn't it? Empire versus independence. Monarchy versus democracy. Checks and balances. A republic of *united* states. And all of that "under

God," meaning we look to something higher, better, more compassionate yet more unforgiving than ourselves. We value our Constitution so much, it remains the touchstone for every law in the land. The document is 250 years old, yet it remains our beating heart. It's stronger than partisan politics; that's the idea.

And the Declaration of Independence that started off this gallant experiment is perhaps the most important mission/values statement ever written, in terms of impact on the world.

Those in power right now seem to believe that partisanship is more important than Constitutional law or Declaration of Independence values. Such partisanship began with James Madison's and Alexander Hamilton's forming parties (and the press that followed) based on personal and political enmity. But partisanship can often clash with our values.

I speak of such values with reverence because these things are sacred and inviolable. How can I—or any American—argue there are some values more in line with the universe than others, more right and true and pure than others? Because some are. Remember the example of deciding whether to sit at the bedside of your dying child or finish that paperwork for your boss? The instinct that compels us to do the right thing, the inherent sense of what's right and less right, resides in all of us. It's what founded our country, what won us the two worst wars in modern history, what sustains the largest, richest, most powerful democracy in the history of planet Earth, despite vast distances both geographic and cultural, both ethnic and religious.

It felt intuitively right to most stakeholders and members of the public not to build on the footprints of the fallen towers. It was simply the right thing to do, and no one needed to explain that to most of us. Ground Zero Values.

Paradoxically, our intuition about what's wrong and right is what keeps us from another civil war like the one that nearly rent us in twain forever. We might disagree with each other in manifold and critical ways, but the vast majority of us remember we're all American, our states are united, we stand for something. If nothing else, it's the relatively peaceful transition of power from one end of the political continuum to the other—or off the whole damn chart entirely for a time. It's what keeps us from clubbing each other in the street. Rabid Democrats and ferocious Republicans, for the most part, don't come to physical blows no matter how heated their rhetoric gets. More of that, please. More awareness of, and clarity about, those things that bond and fuse us rather than fracture and break us. Those Ground Zero Values: a love of liberty and freedom and individual expression. A pride for our stunning endowments and triumphs, even as we keep an honest eye on how far we still need to go.

Stephen R. Covey provides a meaningful example of such true north principles in his book *The Seven Habits of Highly Effective People*. He finds that if you take a group in any auditorium and ask them to close their eyes and point north, everyone points in different directions. That doesn't mean there isn't a true north. North is north by universal law. We can't democratically vote by consensus or Electoral College for south to be north, nor could we tolerate a strongman's dictating that south is north. We cannot wish or will or engineer or force the poles to reverse. In the same way, certain values, certain principles, certain missions, certain goals are more aligned with the universal law, with common sense, humanity, evolution, enlightenment, and God's will. It turns out the core American values we share—or should share—are all aligned with North Star principles. We need only look up at 1 WTC to recall the values responsible for that building—and not a pit and pile—being there.

We Americans believe so strongly in the principles of justice and fairness that they're baked into our founding documents. That doesn't mean the road there isn't littered with major obstacles. But when we keep our collective eye on that destination, we inch closer. The Reverend Doctor Martin Luther King Jr. said, "Somehow the arc of the moral universe is long, but it bends towards justice."[32] Amen.

So, too, America's moral arc is long. I don't agree with the über-conservative ideology that only God can bend that arc, that we foolish mortals confuse ourselves with the Divine if we try to intervene.[33] Of course I don't believe that. Yes, we need God's guidance, but We, the

People, must all endeavor to keep our eyes on the compass, which does not lie, and keep our own footing firmly aimed toward the true north principles on which our great nation was founded. We need to stay as grounded as we were in the days we stood at Ground Zero. We cannot think of King's "somehow" as an abstract concept, dependent on the caprice of otherworldly forces. God has endowed us with the will to, and the means to, rule our own lives. A general does not have to whisper to and nudge every soldier in every trench and battlefield. All good soldiers know the general's will, his or her commands, the mission.

How do we *invest* in the noble values and mission-based actions in praxis? We see it every day. Remember, the concept is easy to grasp: goals come from missions, which begin in values.

Take public transport. The goal of a train is to get passengers from point A to point B owing to some predetermined exigency, such as the need to get from the suburbs to the city for work. But the mission behind that is to do so safely and efficiently, to keep the economy ticking. Maybe there are other sub-missions, too, such as keeping cars off the road to cut down on traffic and pollution, or to provide the economically disadvantaged a cheaper form of transport and expand their work possibilities. All of that stems from values. We're not buying trains per se. We're investing in our values. We value protecting and serving the public. We value the environment. We value growth. We value work. We value our people.

I used the word "exigency," meaning necessity is the usual mother of invention. But remember the problem with that kind of thinking. We can identify it in physical stuff like our infrastructure. We usually wait for a problem to really become a problem before we do anything about it. Then we slap a bandage on it and hope that it holds. When whatever original wound we had gets infected because the bandage is not enough, we put another bandage on the bandage in a patchwork fashion. Maybe we'll give an antibiotic to cut down the symptoms, ease the suffering. But we need to cure problems—or better yet, we need to prevent them in the first place. We need maintenance plans in place that keep things humming so they *don't* go kaput.

Or we might all agree that some big infrastructure job needs doing—say, retooling an airport. But we wind up spending billions and inconveniencing the public for the wrong reasons. We do so in the name of *buying* new, shiny things (with other people's money, it's easy). Or paying for a toy for some politician's egoistic legacy. Investments have value beyond the material. They're necessary and worth the effort, such as increasing capacity. That's the difference between Andrew Cuomo's and George Pataki's agendas.

Sometimes we don't really need a particular infrastructure rebuilt—a new bridge, tunnel, or station—but simply need better maintenance of our existing assets. Commuters and experts almost universally agree that the FDR is among the most derelict and deficient highways in America. Do we need a new East Side artery? Maybe. But maybe we just need to invest more resources into keeping the one we've got working optimally, which it certainly is not. We conserve resources this way. The efforts become more valuable.

Inside the Belly of the Beast

When we do build from scratch, we must build big. If you haven't seen Santiago Calatrava's WTC Transportation Hub and PATH station, I'd highly recommend taking a trip down there. It's an intensely gorgeous and extremely effective station. Our first mission with that was security and safety. Our second, capacity and efficiency. Then, given those values, we let a master architect build one of the world's most beautiful and unique structures:

> *300 pieces of steel erected in a rare, Vierendeel Truss design that features a pair of 150-foot-high "wings" suspended over a glass and steel "body"— built from two sets of specially-designed arches, each weighing between 10-25 tons and standing over 30 feet tall—that allows natural light to penetrate to the rail platforms more than 60 feet below street level.*[34]

Since May 2016, the 350-foot Dey Street Concourse pedestrian tunnel has hooked the Transportation Hub to the already superbly well-organized and exquisite $1.4 billion Fulton Center Station. The *Times* says the former "spaghetti-bowl tangle of dark corridors and baffling signage" has given way to a "kind of Crystal Palace, crowned by a dome that funnels daylight two stories below ground."[35] Both architects, Calatrava and Nicholas Grimshaw, literally used light as a building material. Together those stations will go down in history along with Grand Central, King's Cross, Union Station, and Moscow's Komsomolskaya Metro station.

Here's the thing, though. Those old projects cost more than mere money. They cost blood. Countless men died during their construction. Exploding boilers. Rock collapse. Floods.

Think long and hard about the ground on which you're trekking. You're probably standing on the shoulders of giants—your ancestors, your fellow Americans who died for your freedom, those who voluntarily or unwittingly sacrificed themselves to make the world better, safer, more productive. When you drive over the Brooklyn Bridge, remember the twenty people who died during its erection. Or the ninety-six people (the official count) who died building the Hoover Dam. If you're lucky or blessed enough as I have been several times to travel through the Panama Canal, consider the thirty thousand workers who perished in its making—nearly all of them for completely avoidable reasons—sunstroke, snakebite, beriberi, yellow fever, and the constant enemy, malaria (if only Bill Gates had been around!). Jules Dingler, the canal project's director general of works and chief engineer, saw half his engineers and his entire family die there. After the last funeral, he took all his imported horses, including his own prized stallion, up in into a ravine, shot them all, and left for France.[36]

Nothing had been learned from Suez. A staggering 120,000 construction workers died at the site of the Suez Canal—nearly 6,000 from construction injuries, and up to 25,000 from sheer cold, starvation, and physical exhaustion. Ten died and 145 were injured in a small Suez Canal expansion project in 2014–2015. Would you believe 107,000 people died building two of the world's largest rail systems, and a full thousand died constructing the Erie Canal?[37]

Considered, euphemistically, "industrial fatalities," these happened to real, living people who literally died for our convenience, more often than not in awful ways, under deplorable conditions of disease and danger. That kind of blood investment cannot be underestimated, shouldn't be ignored. Those completed projects stand today as memorials to the people who died in their making.

We've come a long way in terms of safety conditions on worksites. Still, we lost an unacceptable two workers while rebuilding the WTC site. The first, in 2003, was a young contractor, thirty-six, who was crushed by an aerial lift while painting a steel beam. Later, in 2004, a twenty-eight-year-old apprentice carpenter fell and died from the 21st floor.[38] There were thirty-four serious injuries, some of which severely altered those workers' and their families' lives.

And of course there were the three thousand who died in the terror attacks of September 11, 2001. The memory of their sacrifice isn't only memorialized in the footprints of the Twin Towers, but also spiritually imprinted on all of us who had anything to do with transforming that disaster scene into the resplendent site we see today. The WTC site is sacred space. And so is your life, made sacred by those who suffered so you can live.

I didn't intend to end this chapter on such a sobering note, but I'm stunned by this reality regarding investments, as reported in the *Times*:

> *Al Qaeda spent roughly half a million dollars to destroy the World Trade Center and cripple the Pentagon. What has been the cost to the United States? In a survey of estimates by* The New York Times, *the answer is $3.3 trillion, or about $7 million for every dollar Al Qaeda spent planning and executing the attacks. While not all of the costs have been borne by the government—and some are still to come—this total equals one-fifth of the current national debt.*[39]

People, we absolutely have to win this war.

6

Adapt and Compromise

The Green Ledger

We exceeded the original budget for Calatrava's Transportation Hub at the WTC site by nearly 50 percent. We bought a vision and a plan that we estimated would cost $2 billion but wound up costing $4 billion. The decision to devote eight acres, half the site, to the Memorial Plaza honoring the 2,753 souls who died there was the first reason. But we also made a political calculation about ensuring that Memorial Plaza would open by the ten-year anniversary of 9/11.

Fulfilling the mission of building a world-class, masterpiece transportation hub which downtown Manhattan had never seen was never going to be cheap, given the complexities of the site. Of course there were unforeseen financial snags and snowballs. But we'd built into the initial plans enough of a Murphy's Law factor to account for most of that. No, the main reason it cost $4 billion to build that magnificent site—with its two rows of soaring white ribs, open white floor plan and marble mezzanines, peninsular viewing platforms, and skylight ribbon of one thousand operable panes of bomb-resistant Viennese glass—is because we had to adapt to facts on the ground as we went.

Despite the name of this book, we had to build it from the top down. Even though everything was predesigned and the groundwork had been laid for all the nine buildings and associated projects on the site, getting the memorial done meant that first Steven Plate, the Port Authority's director of construction for the site, had to figure out how to build the plaza, fully functioning, in time for the ten-year anniversary, September 11, 2011, and then build *down* so that everything below—the entire Transportation Hub and associated subterranean passages under the whole site—all hooked up properly and operated in concert.

It's obviously cheaper and easier to build from the ground up. But the cost—in terms of public relations—of not opening the memorial on time would have far exceeded the extra capital required. We simply couldn't build one facet of the site, such as the Memorial Plaza or the Oculus (the nickname for architect Calatrava's soaring dove of glass and steel), without it affecting all the others. We faced numerous domino-effect challenges like this. The footings and slurry walls of the original towers had to form the lungs of the underground museum. Parts of the original towers had to be preserved near the PATH lines (you can see the artifacts deep below curved glass cut into the PATH platform).

And then there were the trains, which all had to get up and running as soon as possible and remain accessible throughout the rebuild: the New York City Subway 2 and 3 trains at Park Place; the A, C, and E trains at Chambers Street-World Trade Center; the N, R, and W trains at Cortlandt Street; as well the Newark-World Trade Center line at all times and the Hoboken-World Trade Center line on weekdays.

We restored PATH service between New York and New Jersey by 2003, which no one thought was even possible. All those lines are accessible today on four gleaming white marble train platforms, serving fourteen million passengers a year. A third of a million square feet of prime underground retail space hosts millions more people each year. There are underground passages, including the west concourse connecting the PATH train from New Jersey to the east-west passageway to Brookfield Place. There's a new office/retail complex across from the WTC site. And of course there's a complex operations and security control center, one of the most sophisticated and secret in the world. (Smile, you're on camera!)

During construction, we had to keep all those trains running, the pedestrian traffic (relatively) unimpeded, the West Side Highway moving, and the 315 miles of Hudson River—gazillions of gallons of water—out of our site. British engineers call it a knock-on effect. We think of it as a causal chain of interaction, interpenetration, and inter-relation. Think of the phases and components of a construction project as contagious. You can't futz with one thing without it affecting all the others. That's just the way engineering—and life—works. You're constantly adapting one thing to accommodate another.

Still, could you call the Transportation Hub an excessive waste? Some do.

I grant that $4 billion is a lot of public funding. But the funny thing is, now that 1 WTC, the Memorial Plaza and museum, and the Transportation Hub have all been built—now that people can marvel at the physical manifestation of architects' dreams and engineers' and builders' labor—the blather about boondoggle has died down to barely a whisper.

Grand projects go over budget. It happens. It happened during the U.S. Capitol renovation, the opening of which occurred ten years ago, eight years after groundbreaking, at more than double the $265 million estimated budget.[1]

The George Washington Bridge would have gone substantially over its $60 million budget had it not opened well ahead of schedule in 1931 as the longest suspension bridge in the world—with its steel latticework naked of the concrete envelope faced with pink granite, Beaux Arts design flourishes, grand plazas, and statuary originally proposed by Cass Gilbert of Woolworth Building fame.[2] It is the busiest bridge in the world today, with more than one hundred million vehicles crossing it every year. It's currently undergoing a $1.9 billion, decade-long "Restoring the George" structural health improvement.

Central Park was so over budget the project manager and politicians whose asses were on the line for the expense seriously considered scaling it back to half its proposed size. Can you imagine? There would probably be South Harlem projects there now. Instead, a hundred years later, those cost overages are a postscript to the volumes of praise for the feats of engineering and the stunning results that are front and center in history, and that still endure today. Adapt and become a modernist gem, or die in an unmarked tomb.

For a more contemporary taste of the inescapable snowball effect that causes such sobbing that ultimately signifies nothing, consider that between them, the Big Dig, the Channel Tunnel, and China's Three Gorges Dam went $50 billion-plus over budget.[3]

Again, cost overruns are to be expected, and their time in the salacious headlines will be brief as long as the final product wows in utility, capacity, and, yes, visual appeal. Those paperweights or staplers that occupied such an iconic place in the Financial District of New York City for thirty years were hideously ugly by nearly everyone's standards. Can you imagine the outcry if there had been just *one* of them built, how utterly boring and unattractive that only child would have been? In any case, it wasn't until the twins were snatched from our skyline that anyone really appreciated them.

Construction estimating is just that—estimating. How close can you really get when you're planning ten million square feet of office space in nine towers, six hundred thousand square feet of retail space, and a half-million-square-foot hotel? Remember, too, that Silverstein's insurance payout demanded he rebuild *all* the lost square footage, but we had to do it in *half* the space to accommodate the Memorial Plaza.

Want to avoid costly overruns? It's easy. Calculate every conceivable cost and contingency of your project, budget liberally, allow at least a 30 percent fudge factor for unforeseen (inevitable) emergencies, then double the whole thing. It's a good bet that every project in life—every one—will take twice as long, be twice as hard, and cost twice as much as originally intended.

How do you handle costing a personal improvement project? Remember, costs aren't just financial. Other resources are at stake, not the least of which are time and energy. So what's in your budget? In other words, what will it cost to build the kind of life you really want?

Calculate what relationships it could cost, how much time you should allocate, and what sacrifices you might have to make. Remember that those will only *feel* like sacrifices until you reconcile your values with your behaviors. Once you do that—once you truly decide it's important to your life mission to provide more value at work, for example, you won't miss playing tennis quite as much—it won't feel like a compromise as you adapt along the lines of your values. From a cost-benefit analysis standpoint, where you want to be will cost far less than where you are, given that the most significant "cost" here is your self-worth and sense of purpose.

But don't forget, whatever you're planning it to cost, it'll probably cost double that.

"Okay, Houston, We've Had a Problem Here"

Most of us have seen *Apollo 13*, the dramatization of a malfunction in space caused by an explosion that tore open an oxygen tank in a service module on an April evening in 1970, two hundred thousand miles from Earth.

"There's one whole side of the spacecraft that's missing," Captain Jim Lovell told mission control at the Manned Spacecraft Center in Houston—now the Johnson Space Center—after the rupture gutted the cryogenic tank and nearly consumed the whole ship.[4] In an instant, the moon-bound astronauts—who would have been the third crew to make a lunar landing—found themselves no longer en route to the moon's Fra Mauro Highlands. Instead, they were floating in a cold, dark, and epic odyssey in hopes of somehow returning to terra firma (well, technically, the Pacific Ocean).

What followed was an almost unbelievable coup of innovative engineering and execution by the three astronauts, multiple flight-control teams, thousands of backroom flight engineers and support staff across the country, as well as the spacecraft manufacturer. Those guys seemed destined to plunge headlong into the incinerating atmosphere five weeks later.[5] Instead, they splashed down safely a few days after the explosion. The whole world was watching, including my wife, my kids, and me.

How did they do it? With a combined technology far less impressive than your average Bluetooth-enabled microwave oven. All the main factors were foundational. Clarity of mission. Respect for authority (in this case, for the flight director Gene Kranz's total license). Flawless teamwork coupled with individual responsibility. The segregation of paralyzing emotion from practical mission. And finally, perhaps most important, endless drilling and paramilitary-level preparation.

This was the same way the first responders managed to stay so calm during their rescue and recovery efforts at Ground Zero. If you simulate an activity enough times, a kind of muscle memory takes over. Success or failure of any endeavor, no matter how difficult, can be predicted with near precision based on the relative foundation of training, preparation, and contingency planning. Adapting doesn't feel like adapting while you're doing it—it just feels like acting in accordance with an evolving situation.

As controllers clambered to track down the source of the outgassing Lovell reported seeing, Kranz broadcast to everyone online, "Okay, let's everybody keep cool.... Let's solve the problem, but let's not make it any worse by guessing."[6] Krantz knew that one false move could kill the astronauts, not to mention destroy NASA's reputation forever, ensuring its defunding. But he also knew that in case of dire emergency, they had a key resource at their disposal: an intact lunar module attached to the Odyssey, which the crew could use to sustain life a little longer while mission control kept trying to solve the problem. He also had, serendipitously, a super-trained command module pilot, Ken Mattingly, whom NASA had scrubbed from the mission at the eleventh hour because he had been exposed to German measles. Mattingly stayed awake for days running sims in a training module, which contributed substantively to the rescue mission.

Planning for contingencies means no failures of imagination. If you can think it up, it can happen. Even if you can't imagine it, it can happen. In my eighty-year-plus lifespan, our society has progressed from the invention of electronic television to virtual reality. In the same way you can plan for the worst-case scenario (squirreling away funds for a rainy day, insuring yourself and your family, investing, getting a go-bag ready both literally and figuratively for another 9/11), you can design your dream life. Put it on paper. What would you dare to dream if failing were impossible? Now how do you get from here to there? Not just to the next stop on the journey—the distance you can see in your headlights—but to your destination in all its future splendor.

I recommend you plan attitudinally for your ideal life, top down. In other words, start with the end in mind. If you feel called to become a priest, start acting like a priest. Act as if you have already reached your destination. I know this might seem paradoxical, even illogical. Shouldn't you go one step at a time? Sure. You're going to go one step at a time. But you'll tend to leap farther if you act as if you're already there.

Yes, you have to take away one five-pound bucket of shit at a time. Then you have to put up one floor at a time. But if all you're doing is slogging ahead one exhausting step at a time, you risk giving up entirely. It'll seem too hard, too distant, too mystifying.

This is one benefit of being an engineer: you've always got the plans in front of you so you know what you're building, even as you must discover how to make it along the way. Climbers tackling Mount Everest have to take one tiny, tiring, breath-robbing step at a time, yes. But how many do you think aren't imagining the summit? I bet every one of those adventurers focuses at the same time on the individual steps *and* the motivational peak. Start with the notion "All right, thank God we're back on Earth; we landed safely," then reverse-engineer it. "How the hell did we get here?"

Along the way, though, plan for pitfalls. Train. Learn from your mistakes and the errors of those who've preceded you on your path. We built 1 WTC with a near impenetrable steel-and-concrete core as strong as the outside walls at the foundation—stronger than anything ever built. There's a haven there, a building within a building, if anyone ever needs to escape a disaster. There are dedicated stairs for first responders, and for tenants and visitors, so there will never be a flow/counterflow problem—a major lesson we learned from the 9/11 disaster.

What's your inner core made of? On what unassailable values is it based? We all have the ostensibly superhuman capacity to survive terrible things, to make it over, or through, monstrous obstacles. We can all be like Elie Wiesel, Viktor Frankl, Mother Teresa, Frederick Douglass, Crazy Horse, and Abraham Lincoln, whose strength of character and faith shielded them through their various trails of tears and hikes in hell. The spirit is strong even when the flesh is weak—that's how we're built. The Apollo 13 astronauts shivered and puked the whole time. But they didn't give up.

I've got a great story about a person a lot like that. Her name is Cathy Blaney. She works for Michael Bloomberg and helped him with the museum at Ground Zero. Back in the day, she worked for me. Before the Port Authority, she was an aide-de-camp of sorts during my tenure as ambassador in Trinidad and Tobago. My friend Tony Bevilacqua was getting ordained as a cardinal in Rome, and I was there. I called my assistant, Cathy, from Rome to check on things on the home front in Port of Spain, and there was something about her voice…

"Are you okay?" I asked her.

"I'm fine," she said. "Everything's fine." No woman has ever uttered those words without her internal world in flames.

In this case, it was a car accident. I had a state car with a driver in Trinidad. He was driving it and she was following him in my car, taking it from the embassy to my residence, about a mile away in the hills. The driver could hit a button to drop huge steel barriers in case of peril— an assassination attempt or whatever. Without thinking, as soon as he passed the threshold, he hit that button. He forgot about Cathy. She slammed into the barricade and went straight through the windshield. My jaw dropped as she told the story.

"Oh, my God. I wrecked your car, Charlie."

"The car? Screw the car. Are you all right?"

"Yeah, don't worry about it. So listen, about that dispatch to the Gipper…"

The woman was a machine. I would say it's good to know the difference between a scraped knee and serious internal injuries and act accordingly. But the poor woman went flying through a Lincoln Town Car windshield, tumbled over pavement, got up, adjusted her skirt, and just kept working.

Maybe I'd modeled that. I'd told her this story: It was 1971 or '72. We were twenty-eight feet below Sunrise Highway in the middle of the Massapequa Preserve, having lowered the water table about twenty-two feet. We were running two levels of huge dewatering pumps at well points every five feet, pulling millions of gallons up and out to storm drains. One night, the pressure of our building was just too much. Massapequa Lake overflowed and filled our trench and cut off all the pumps in our excavated sewer line. The next morning, Mario Posillico's brothers, Dominic and Joe, were losing their minds, up to their asses in eels. Eels were jamming the pumps. "What the hell are we going to do now?"

I said, "You fucking guys get out of here. I'll take care of it. I'm in charge. Stop complaining." Twenty-four hours and many more pumps, mechanics, and stages later, the catastrophe was averted and we were back on track.

Not every catastrophe can be averted with so little loss. When I was ambassador, in July 1990, a Friday afternoon, local Muslim terrorists instigated a major coup against the government of Trinidad and Tobago. Forty-two insurgents stormed the Red House, the nation's Parliament, taking the prime minister and most of his cabinet hostage. They brutalized the PM and shot his defense minister. Simultaneously, another seventy-two insurgents attacked the offices of Trinidad and Tobago Television. They assaulted the embassy, too, of course. The damn CIA, as I had predicted, had been entirely useless.

Extensive looting and arson ensued—most of downtown burned. There were millions lost in property damage. Twenty-four people died. In the end, a brave colonel who was in charge of the army down there gave the hostage-takers an ultimatum: surrender or die. The PM was also a hero. These guys kept their cool and saved their nation.

I'll give you another great example of adapting on the fly and using values as a touchstone. The mayor of Sant'Angelo dei Lombardi, Rosanna Repole, was a young schoolteacher in 1980 when her father, the mayor, along with the captain of the carabinieri, the town priest, and many other town officials, were killed in the earthquake. The town tapped her as mayor, a role she reluctantly took on. She dedicated her entire life to her fellow Santangelesi, who became her adopted family. She rose to become president of Avellino province and was recently offered the role in Rome of national infrastructure minister, based primarily on her dogged rebuilding of her ruined village.

Problem Child

Those who survive have a built-in, or hard-won, ability to bounce back from challenges. We faced innumerable problems in the construction of the new WTC. Every single day in the early years post-9/11 was an exercise in adaptability. Every single day. I'd go to a community listening session thinking I knew what to expect, and something would come out of left field. My guy, Glenn Guzi, always knew we should never come from a place of "no," but instead from a place of "let's see if that makes sense and if that's doable." That's adaptability.

Almost three thousand people were killed and thousands more weren't able to live in their homes anymore. Businesses were disrupted. The whole world had changed politically. "Adaptability and flexibility—that's the only way you can survive and think when the stakes are that high," says Guzi. "If you can't be open-minded to thoughts beyond your own, you'll never be able to get a project like that done."

That's usually easier said than done. I'll tell you just one story. Unless you were an insider, you were probably one of the hundreds of millions of people who complained about how long it took to rebuild after the attacks. I don't blame you. I, too, wish it had happened faster. But you have to understand the enormous challenges we faced. On July 4, 2004, after years of planning, we held a public groundbreaking led by Governor Pataki, Mayor Bloomberg, and New Jersey governor James McGreevey.[7] That Independence Day, we dedicated the "Freedom Tower" corner-stone, a twenty-ton block of Adirondack granite with these words:

To honor and remember those who lost their
lives on September 11, 2001 and as a tribute to the
enduring spirit of freedom

They were chiseled upon silver in the New Yorker-designed Gotham typeface. The original plan was to include more words taken from a Pataki speech that May. But Pataki said, "Less is more." He was adamant that we remove his name from the inscription, along with McGreevey's and Bloomberg's. That's Pataki.

We went straight back to planning, surveying, and engineering at the site. Lots of compromising and adapting took place in that period on several fronts. In the original plans, the tower was set twenty-five feet from West Street. The Port Authority, as the site owner, and the LMDC, as the rebuilding director, had each consulted with third-party security firms to assess the public safety of the proposed site plan. New York City police commissioner Raymond Kelly had seen our threat-assessment risk analyses through 2004 and '05, and we had the NYPD's sign-off.

By April 2005 we'd issued the equivalent of a conditional building permit to Larry Silverstein. He, too, had been meeting regularly with Kelly's deputy commissioner for counterterrorism, Michael Sheehan, and the NYPD about the plans. Everyone knew the plans.

Then Kelly wrote a letter. He called for a meeting. Mayor Bloomberg, Governor Pataki, I, and a couple of others heard him out. At the meeting, he said, "I've been talking to my security people, Washington, and [Mayor] Bloomberg. We've been analyzing the plans. We're going to have to ask you guys to move the tower east." He wanted us to move it sixty-plus feet. They would ultimately ask for a tower setback up to a hundred feet away.

Kelly claimed Deputy Sheehan had notified us already, saying he was worried about the "insufficient standoff distance" of the Freedom Tower to West Street. They were concerned about the proposed glass fairly low down on the tower and the thousands of unchecked eighteen-wheelers that would rumble by every day. The setback and other design changes could "harden" the building against a major blast."[8]

It was the first I'd heard of it. It made no sense. Why on earth would we have spent all those months working up plans and preparing the site if we knew it was a no-go? More important, why the hell would Kelly and his crew let us go on and on with that work—hundreds of thousands of hours, all told—if they'd known for months the building would have to be redesigned?

I was livid beyond the beyonds. In that one fell swoop, Kelly had just set us back years in terms of public trust. This all could have been avoided. Not to mention that he rendered the Freedom Tower uninsurable, scaring away all potential tenants. I jumped up. "Holy Christmas!" I yelled. Pataki took my arm, said calmly, "Mister Chairman, relax. It's going to be okay. This is just a hiccup." But privately the governor was equally dismayed, and publicly embarrassed.[9] Even Silverstein—who'd specifically asked for the new NYPD risk assessment—was enraged.

These people were sitting around all that time while we were planning, getting ready to pour foundations. Governor Pataki and I had stood onstage in front of the whole world and dedicated the cornerstone. Now, *bam*! A two-year delay. A twenty-ton slab of granite had to be jimmied up and slinked off to a warehouse on Long Island with less than zero fanfare. There was intense public outcry.

Sounds like a small thing, shifting things a few dozen feet? Think again. Everything from the seventy-foot-deep foundation to the first one hundred vertical feet needed to be reconsidered, re-engineered, rethought. No one was happy. "I don't want to say the police have been irresponsible, but where were they until this month?" said Whitehead, chair of the LMDC. "I wish they had called attention to the seriousness of the problems earlier, rather than at this late stage."[10]

In hindsight, I understand it was a necessary adjustment, which ultimately made our tenants and visitors safer. But I still sometimes pace off those added feet from the West Side Highway to the present location of 1 WTC and mourn the lost time. As an old surveyor and engineer, I like to get things done and done right—the first time. But Pataki was right in counseling me privately that the perfect is the enemy of the good.

It's true. It wasn't until April 27, 2006, that we all gathered again and rededicated the groundbreaking. Pataki looked at me gently as I slowly shook my head onstage. His trademark smirk told me everything I needed to know—*See, just a hiccup.*

The perfect is the enemy of the good. None of us will ever be done with perfecting ourselves—we'll forever be trying to complete our personal punch lists. As we tick off one item, another will surely pop up. An honest inventory of our design flaws never hurts. You know we built 1 WTC as the world's safest skyscraper. Features like the dam-dense concrete, rebar as thick as your arm, and steel walls and core make it nearly impenetrable to blasts—*nearly*. But cockiness can get people killed. I would never say—nobody did or should—"Even God can't sink this ship." We'd be asking for trouble.

The same is true in your personal re-engineering project. Perfection is a worthwhile but ultimately impossible goal. We're all works in progress. We're all sinners. The major difference between us is that some of us have capitulated to our imperfection, given up on self-improvement. Or worse, given in to our baser instincts such as greed and pettiness, power and avarice (just study the sordid life of Sheldon Silver or watch a Trump press conference). Others of us continuously strive to improve, to learn from our mistakes. Half of our marriages end in divorce, mine included. We try to do it better the next time.

Do you have a higher power? Statistically speaking, more and more of us do not. How sad that is. Atheism is alarmingly on the rise. Church attendance is down. When I was a kid, people went to church every Sunday, even with the flu. In fact, I went on Sundays *and* Wednesdays for an hour. Church attendance was the barometer of spiritual health.

Do you have to go to church to be a good person? Of course not. I've occasionally lapsed myself in that duty. But ask yourself to what you are devoted. Is it TV? The internet? Booze? Personal ambition? Take a look beyond your words and even your beliefs to your deeds and works.

As you sow, so shall ye reap. This counsel is very powerful. It's similar to the Eastern concept of karma, which is not at all incompatible with Christian doctrine. In fact, it's pretty much the same thing worded differently. I like the idea of a Master Gardener advising us on our spiritual space—what to plant and how to tend our lives.

My ideal day now is getting up early, fast-walking for an hour in Central Park, and having a healthy breakfast at the Loeb Boathouse, followed by a productive day writing or working on philanthropic projects, then finishing up with a good dinner among friends and family.

It's important to leave a legacy. In Central Park, there's a Walk of Fame next to the Japanese garden I sometimes pass. There are stones laid into the new walk with names of those who grew up in Brooklyn and later became famous. I don't know why he did it, but in 2002, the Brooklyn borough president honored me by dedicating a stone with my name next to Susan Hayward's and Lauren Bacall's. The ceremony was grand. Seeing that stone from time to time reminds me to try to do something for someone else every day.

That could mean organizing a concert at Madison Square Garden with Frankie Avalon, Frankie Valli, Tom Dreesen (who opened for Frank Sinatra for thirteen years), and others to raise money for Sant'Angelo dei Lombardi. I continue to help in that community, building schools, athletic fields, and nursing homes. We recently raised a replacement steeple on my family's ancestral cathedral, which took us twenty-two years to rebuild. You can see it for miles around, and the night the bells rang out again for the first time since the earthquake was one of the most emotional ones of my life.

The Church of St. Anthony's Martyr is the cathedral of the archdiocese of Sant'Angelo dei Lombardi-Conza-Nusco-Bisaccia. It dates from the eleventh century—the crypts of the bishops in the basement remain intact. Though built in the Norman period, it was originally Byzantine style, but was adapted to Baroque by the bishop after an earthquake in 1664. It was built in the traditional cruciform, with three naves, a travertine façade with three reliefs of Christ the Redeemer and the archangel Michael (the town's patron saint), and a stone campanile. My father used to climb that four-hundred-year-old bell tower when he was a boy. He and his friends would swing from the ropes, ring the bells, and confuse the citizens.

All of it was a pile of stones after November 23, 1980. We had to rebuild it in three stages, injecting concrete into the walls and wrapping the whole thing with vertical and horizontal stabilizing bars of steel. Beams shoot through the interior, too, making it a model of reconstructed church architecture for all of Italy. This is good, given that it sits directly on a vertebra of the shaky spine of the Apennine Peninsula, the Irpinia Red Zone.

"We don't know how to defend ourselves a hundred percent" from another earthquake, says chief engineer Michele Candela. But we've done the best we can do.[11]

I'm so proud to have been responsible for that reconstruction. There's going to be a PBS documentary, *From the Ground Up*, a kind of modern version of *A Bell for Adano*.

It could have all been different. After we made that fateful first trip post-Irpinia in 1980, after "Okay, boys, the Holy Father wants to see you," you know how I wound up with a position in the Reagan administration? We all went out to dinner after leaving the Vatican. John Volpe, the former governor of Massachusetts, leaned over his cavatelli and said, "Charles, I've been watching you this past week. You should be in politics. I'm on the transition committee with President Reagan. I'd like to sponsor you to run against Mario Cuomo. We have to unseat that bastard."

And I said, "Governor, thank you, I'm flattered, but I'll tell you something. I'm not in politics. I don't want to be in politics. My partner, Mario Posillico—that's your guy. He loves politics and he's dealing with the politicians all the time. I don't."

"Well, send me your résumé anyway," he said.

Together we found an upstart state legislator, the former mayor of Peekskill. An Italian, Irish, and Hungarian boy. His name was George Elmer Pataki. He was our guy. But John mentioned me to Reagan, and Reagan came knocking.

Politics per se wasn't for me. But I urge you to serve as I did. Don't only vote. Run for office. Build something. Leave a legacy. Remember the long path. What can we do for our grandchildren's country?

Right around the corner from my childhood house was Prospect Park. We never let a nice day pass without being out there playing baseball, boating or fishing on the lake, horseback riding, sledding, playing hide and seek, picnicking. We had access to the zoo, the Brooklyn Botanic Garden. My God—it was a paradise. And Ebbets Field, where my beloved Brooklyn Dodgers played. I went to many games there with friends. We all belonged to the "Knothole Gang," they called it. The Dodgers gave more than two million free passes to kids during the 1940s and '50s. We'd lie on the pavement to grab a view from under the center-field gate at Ebbets. Tickets were sixty cents.

That's what I wish for young Americans. We're on our way. New York state comptroller Thomas DiNapoli reported in September 2016 that the Lower Manhattan area had a more diverse and faster-growing economy than the city's average figures. In the previous fifteen years, the general population of the area around the former WTC had more than doubled, and the population of children had tripled, according to DiNapoli's report, while the city's overall population had grown only 4.3 percent.[12] It's all on the upswing.

If we keep the right attitude, we'll keep growing. That's the real secret of our success.

Acknowledgments

I want to thank my wonderful partner for more than ten years, Marilyn Alfeld, for her support and friendship; and my loving family, Larry, Carla, and Prudence who read through some sensitive issues and suggested minor adjustments.

I also thank the former N.Y. governor, George E. Pataki, for years of trust and friendship.

My attorney, Joseph A. Edgar, who deftly handled the legal affairs related to publishing.

My collaborator, Ian Blake Newhem, helped bring shape to my message and was an invaluable asset to this book.

Lacy Lynch, Dabney Rice, and the crew at Dupree/Miller & Associates, for their unwavering advocacy.

Anthony Ziccardi, Michael Wilson, Elena Vega, Maddie Sturgeon, and everyone at Post Hill Press for believing in this mission, and for the incredible and thankless work they do. Who knew how complicated it is to get a book on the shelf?

My chiropractor and dear friend, Doctor Michael C. Smatt, for his selfless service at Ground Zero. Doctor Smatt, along with the New York Chiropractic Council, organized several sites so police officers, firefighters, rescue workers, and volunteers could receive free adjustments seven days a week, twenty-four hours a day, for a year.

All the women and men who stepped up on 9/11 in the days, months, and years after, sacrificing greatly for the country we all love.

And, finally, I thank you, the reader, for your patriotism and commitment to returning America to greatness.

Endnotes

Opening Quote

1 Roosevelt, Theodore. 1899. "The Strenuous Life." *In Theodore Roosevelt, The Strenuous Life: Essays and Addresses*. New York: The Century Co. 1901.

Introduction

1 "PlaNYC: A Greener, Greater New York." The City of New York Mayor's Office of Recovery & Resiliency. 2007. http://www.nyc.gov/html/planyc/downloads/pdf/publications/full_report_2007.pdf. Accessed October 17, 2016.

2 Cartwright, Mark. "Aqueduct." *Ancient History Encyclopedia*. September 1, 2012. http://www.ancient.eu/aqueduct/. Accessed February 7, 2017.

3 Du, Lisa and Wei Lu. "U.S. Health-Care System Ranks as One of the Least-Efficient: America is Number 50 out of 55 Countries That Were Assessed." Bloomberg.com. September 28, 2016. https://www.bloomberg.com/news/articles/2016-09-29/u-s-health-care-system-ranks-as-one-of-the-least-efficient. Accessed July 14, 2017.

4 Desilver, Drew. "U.S. Students' Academic Achievement Still Lags That of Their Peers in Many Other Countries." Pew Research Center. February 15, 2017. http://www.pewresearch.org/fact-tank/2017/02/15/u-s-students-internationally-math-science/. Accessed July 14, 2017.

5 "Right Direction or Wrong Track? 36% Say U.S. Heading in Right Direction." *Rasmussen Reports*. July 10, 2017. http://www.rasmussenreports.com/public_content/politics/mood_of_america/right_direction_wrong_track_jul10. Accessed July 14, 2017.

6 Cohen, Arianne. "World Trade Center 7 Report Puts 9/11 Conspiracy Theory to Rest." *Popular Mechanics*. August 20, 2008. http://www.popularmechanics.com/technology/design/a3524/4278874/. Accessed July 1, 2017.

7 Templeton, Tom and Tom Lumley. "9/11 in numbers." *The Guardian* (UK). August 17, 2002. https://www.theguardian.com/world/2002/aug/18/usa.terrorism. Accessed July 13, 2017.

Chapter 1

1   2013 Report Card for America's Infrastructure. Society of Civil Engineers. http://www.infrastructurereportcard.org/bridges/. Accessed October 12, 2016.

2   "AAA Study: Road Rage on the Rise." *CBS Miami.* July 14, 2016. http://miami.cbslocal.com/2016/07/14/aaa-study-road-rage-on-the-rise/. Accessed February 7, 2017.

3   Mclean, Robert. "Americans Were Stuck in Traffic for 8 Billion Hours in 2015." *CNN Money.* March 15, 2016. http://money.cnn.com/2016/03/15/news/us-commutes-traffic-cars/. Accessed February 7, 2017.

4   Stone, Arthur. "A New Look of Commuting in the United States." *The Evidence Base.* May 9, 2016. http://evidencebase.usc.edu/?p=852. Accessed February 7, 2017.

5   Ballmer, Steve. "I Crunched the Numbers of the US Government: Here's What I Learned." *Time* magazine. July 31, 2017.

6   Lieb, David A. "Upkeep of Roads, Bridges Costly." *Associated Press.* May 7, 2017.

7   Braverman, Beth. "10 States with the Worst Highways in America." *The Fiscal Times.* July 15, 2016. http://www.thefiscaltimes.com/Media/Slideshow/2016/07/15/10-States-Worst-Highways-America?page=9. Accessed February 7, 2017.

8   DePillis, Lydia. "Trump unveils infrastructure plan." *CNNMoney.* February 12, 2018. https://money.cnn.com/2018/02/11/news/economy/trump-infrastructure-plan-details/index.html. Accessed November 25, 2018.

9   Malone, David. "Growth Spurt: A Record-Breaking 128 Buildings of 200 Meters or Taller were Completed in 2016." *Building Design & Construction.* January 23, 2017. https://www.bdcnetwork.com/growth-spurt-record-breaking-128-buildings-200-meters-or-taller-were-completed-2016. Accessed July 14, 2017.

10 Badger, Emily and Niraj Choski. "How We Became Bitter Political Enemies." *New York Times.* June 15, 2017. https://www.nytimes.com/2017/06/15/upshot/how-we-became-bitter-political-enemies.html. Accessed July 3, 2017.

11 Getlin, Josh. "In Brooklyn, Plan for Minor League Baseball on Coney Island Assailed." *LA Times.* August 18, 2000.

12 Staba, David. "Cityside: John Hearth Flames on Once Again." *Niagara Reporter.* December 12, 2006. http://niagarafallsreporter.com/citycide12.12.06.html. Accessed October 16, 2016.

13 Stone, Roger. "Pataki and Gargano Under the Microscope." *The Stone Zone.* July 14, 2008. http://www.stonezone.com/article.php?id=106. Accessed October 16, 2016.

14 Hogrefe, Jeffrey. "Dia hired Ed Hayes to 'Deal with the Garganos.'" *NY Observer.* March 22, 1999. http://observer.com/1999/03/dia-hired-ed-hayes-to-deal-with-the-garganos/. Accessed October 16, 2016.

15 Stone. "Pataki and Gargano Under the Microscope."

16 Frantz, Joe Sexton and Douglas. "A Powerful Fund-Raiser Who Also Oversees State Contracts." *The New York Times.* September 3, 1995. https://www.nytimes.com/1995/09/03/nyregion/a-powerful-fund-raiser-who-also-oversees-state-contracts.html. Accessed September 11, 2018.

17 "Charles A. Gargano: Doctor of Commercial Science." *Pace University.* 2005. https://www.pace.edu/sites/default/files/files/hdr-2005-charles-a-gargano.pdf/. Accessed October 16, 2016.

18 Labash, Matt. "Roger Stone, Political Animal, 'Above All, Attack, Attack, Attack—Never Defend.'" *The Weekly Standard*. November 5, 2007. http://www.weeklystandard.com/roger-stone-political-animal/article/15381. Accessed October 16, 2016.

19 Newport, Frank, Lydia Saad, and Michael Traugott. "The More Americans Know Congress, the Worse They Rate It." Gallup poll. September 25, 2015. http://www.gallup.com/poll/185912/americans-know-congress-worse-rate.aspx. Accessed October 12, 2016.

20 Ehley, Brianna. "5 Most Egregious Examples of Government Waste This Year." *The Fiscal Times*. December 30, 2014. http://www.thefiscaltimes.com/2014/12/30/5-Most-Egregious-Examples-Government-Waste-Year. Accessed February 7, 2017.

21 The Port Authority of NY & NJ. *Overview of Facilities & Services*. http://www.panynj.gov/about/facilities-services.html. Accessed July 4, 2017.

22 Massey, Daniel. "New Jersey vs. New York City in Economic Race." *Crains New York*. May 15, 2001. http://www.crainsnewyork.com/article/20110515/FREE/305159969/new-jersey-vs-new-york-city-in-economic-race. Accessed November 14, 2018.

23 "State Population 2017." *2017: World Population Review*. 2017. http://worldpopulationreview.com/states/.

24 "New York City Population 2017," *World Population Review*. 2017. http://worldpopulationreview.com/us-cities/new-york-city-population/.

25 "Gross Domestic Product (GDP) by State." Bureau of Economic Analysis, U.S. Department of Commerce. 2016. www.bea.gov/iTable/.

26 "State Minimum Wages/2017 Minimum Wage by State." National Conference of State Legislatures. 2017. http://www.ncsl.org/research/labor-and-employment/state-minimum-wage-chart.aspx.

27 Murray, Ed. "$15 Minimum Raises Incomes." Op-ed. July 6, 2017.

28 Hartmann, Margaret. "Jesus Christie: New Jersey's Government Shutdown Is Over, But We'll Always Have Beachgate." July 5, 2017. *The Daily Intelligencer*. http://nymag.com/daily/intelligencer/2017/07/new-jersey-government-shutdown-ends-beachgate.html. Accessed 7-20-17.

29 Mcgeehan, Patrick. "Christie Halts Train Tunnel, Citing Its Cost." *The New York Times*. October 07, 2010. https://www.nytimes.com/2010/10/08/nyregion/08tunnel.html. Accessed October 26, 2018.

30 Massey.

31 Smith, Jack. "$37 Screws, a $7,622 Coffee Maker, $640 Toilet Seats: Suppliers to Our Military Just Won't be Oversold." *Los Angeles Times*. July 30, 1986. http://articles.latimes.com/1986-07-30/news/vw-18804_1_nut. Accessed February 5, 2017.

32 "Contract-Awarded Labor Category." General Services Administration. https://calc.gsa.gov/?q=engineer&proposed-price=700&sort=-current_price. Accessed February 7, 2017.

33 Gladstone, Steve. "Procurement and Government Waste". Gladstone 2016. https://gladstone2016.com/procurement-and-government-waste/. Accessed February 5, 2017.

34 Porter, Logan. "Defense Contractors Make Three Times Private Sector Wages, Numbers Show." *Washington Examiner*. February 12, 2014. http://www.washingtonexaminer.com/defense-contractors-make-three-times-private-sector-wages-numbers-show/article/2543940#!. Accessed July 15, 2017.

35   Burnett, H. Sterling. "So Much for Obama's Pledge to Transparency." *Forbes*. March 2, 2016. http://www.forbes.com/sites/realspin/2016/03/02/so-much-for-obamas-pledge-to-transparency/#3d2d3e9a6abc. Accessed February 7, 2017.

36   Johnson, Jenna and Matea Gold. "Trump Calls the Media 'The Enemy of the American People.'" *Washington Post*. February 17, 2017. https://www.washingtonpost.com/news/post-politics/wp/2017/02/17/trump-calls-the-media-the-enemy-of-the-american-people/?utm_term=.0d77e88db976. Accessed February 28, 2017.

37   Edsall, Thomas B. "The War on Entitlements." *New York Times* Opinionator blog. March 6, 2013. https://opinionator.blogs.nytimes.com/2013/03/06/the-war-on-entitlements/?_r=0. Accessed February 5, 2017.

38   Harshbarger, Rebecca. "Nearly All LIRR Workers Have Disability Payments Approved." *New York Post*. August 6, 2014. http://nypost.com/2014/08/06/nearly-all-lirr-workers-have-disability-payments-approved/. Accessed March 1, 2017.

39   Post Editorial Board. "Teachers Union Would Rather Cover Up Abuse than Protect Our Kids." *New York Post*. June 8, 2016. http://nypost.com/2016/06/08/teachers-union-would-rather-cover-up-abuse-than-protect-our-kids/. Accessed July 15, 2017.

40   Brill, Steven. "The Rubber Room." *The New Yorker*. August 31, 2009. Academic OneFile. Accessed July 20, 2017.

41   Edelman, Susan. "City Pays Exiled Teachers to Snooze as 'Rubber Rooms' Return." *New York Post*. January 17, 2016.

42   "Nobel Secretary Regrets Obama Peace Prize." *BBC News*. September 17, 2015. http://www.bbc.com/news/world-europe-34277960. Accessed July 15, 2017.

Chapter 2

1   Peggy Noonan. "A New Kind of 'Credibility' Gap." *Wall Street Journal*. September 20, 2013. https://www.wsj.com/articles/noonan-a-new-kind-of-credibility-gap-13796 26830?tesla=y. Accessed July 24, 2017.

2   Morton, Brian. "Falser Words Were Never Spoken." *The New York Times*. August 29, 2011. https://www.nytimes.com/2011/08/30/opinion/falser-words-were-never-spoken.html?mtrref=www.google.com&gwh=CF015AFFACBB136574A723C-788186B04&gwt=pay&assetType=opinion. Accessed September 12, 2018.

3   Hagger, Nicholas. *The Secret American Dream: The Creation of a New World Order with the Power to Abolish War, Poverty, and Disease*. London: Watkins Publishing. 2011.

4   Bloom, Josh. "Weapons of Ash Destruction: The Chemistry of Kilauea." *American Council on Science and Health*. May 10, 2018. https://www.acsh.org/news/2018/05/10/weapons-ash-destruction-chemistry-kilauea-12937. Accessed September 11, 2018.

5   "Ceremony Closes 'Ground Zero' cleanup." CNN. May 30, 2002. http://edition.cnn.com/2002/US/05/30/rec.wtc.cleanup/. Accessed July 16, 2017.

6   "Fostering Independence in Children." PBS.com. http://www.pbs.org/parents/education/learning-disabilities/fostering-independence-in-children/. Accessed July 16, 2017.

7   Urist, Jacoba. "What the Marshmallow Test Really Teaches About Self-Control." *The Atlantic*. September 24, 2014. https://www.theatlantic.com/health/

archive/2014/09/what-the-marshmallow-test-really-teaches-about-self-control/380673/. Accessed November 14, 2018.

8   O'Reilly, Bill and Martin Dugard. *Killing Lincoln: The Shocking Assassination that Changed America Forever.* New York: Henry Holt. 2011. p. 177.

9   "U.S. National Debt Clock: Real Time." State of Texas Debt Clock. http://www.usdebtclock.org/. Accessed September 11, 2018.

10 Vernon, Steve. "Let's Debunk This Social Security Myth." *Moneywatch.* July 6, 2016. http://www.cbsnews.com/news/lets-debunk-this-social-security-myth/.

11 Social Security Administration. "Summary: Actuarial Status of the Social Security Trust Funds." June 2016. https://www.ssa.gov/policy/trust-funds-summary.html.

12 "Final Reports from the NIST World Trade Center Disaster Investigation." NIST Engineering Laboratory. https://www.nist.gov/engineering-laboratory/final-reports-nist-world-trade-center-disaster-investigation. 2005. Accessed March 10, 2017.

13 "FAQs—NIST WTC Towers Investigation." National Institutes of Standards and Technology, Engineering Laboratory. https://www.nist.gov/el/faqs-nist-wtc-towers-investigation. Accessed July 12, 2017.

14 "FAQs—NIST WTC Towers Investigation." Accessed July 12, 2017.

15 Hanser, David A. *Architecture of France: Reference Guides to National Architecture.* Westport, CT: Greenwood Press. 2006.

16 "11 Billion Gallons of Sandy Sewage Overflow." *Climate Central.* http://www.climatecentral.org/news/11-billion-gallons-of-sewage-overflow-from-hurricane-sandy-15924. Accessed March 9, 2017.

17 Jansen Kraemer Jr., Harry M. *From Values to Action: The Four Principles of Values-Based Leadership.* Jossey-Bass. April 2011.

18 "We Will Rebuild: President Bush, NY Mayor Giuliani and NY Governor Pataki Phone Conversation." *American Rhetoric: Rhetoric of 9-11.* MP3 transcript. September 13, 2001. http://www.americanrhetoric.com/speeches/gwbush911calltonewyork.htm. Accessed June 19, 2017.

19 Meserve, Jeanne and Mallory Simon. "Web Site Posts What It Says Are Half Million Text Messages From 9/11." *CNN.* November 26, 2009. http://edition.cnn.com/2009/US/11/25/september.11.messages/. Accessed July 17, 2017.

20 "The Law of Posteriorities." Brian Tracy's Self Improvement & Professional Development Blog. May 12, 2015. Accessed October 26, 2018. https://www.briantracy.com/blog/leadership-success/the-law-of-posteriorities/.

Chapter 3

1   Buettner, Dan. "Longevity, The Secrets of Long Life." *National Geographic.* November 2005.

2   Levitt, Stephen D. and Stephen J. Dubner. "What Should You Worry About." *Parade. The Washington Post.* October 18, 2009. p. 9.

3   "World on Sunday." *New York World.* April 1906. Cover page.

4   Vespasiano, Michele. Interview with the ghostwriter, March 29, 2017, Sant'Angelo dei Lombardi.

5   Vespasiano. *Assistenza e migrazione tra XIX e XX secolo: Le società di mutuo soccorso di Sant'Angelo dei Lombardi.* Natan Edizione 2017. P 50.

6 History of New York City's Water Supply System. http://www.nyc.gov/html/dep/
html/drinking_water/history.shtml. Accessed April 22, 2017.

7 Marzulli, John and Bill Hutchinson. "Tommy Shots Taught Me How to Kill,
Former Colombo Capo Testifies." *New York Daily News*. March 27, 2012. http://
www.nydailynews.com/new-york/brooklyn/tommy-shots-taught-kill-colombo-ca-
po-testifies-article-1.1051008. Accessed July 17, 2017.

8 Gray, Christopher. "Ghost Buildings of 1929." April 23, 2009. *New York Times*.
http://www.nytimes.com/2009/04/26/realestate/26scapes.html. Accessed April 20,
2017.

9 "Hoover Dam: Frequently Asked Questions and Answers." United States Bureau
of Reclamation. Last updated March 12, 2015. https://www.usbr.gov/lc/hooverdam/
faqs/damfaqs.html. Accessed March 1, 2017.

10 Stockton, Nick. "It Took 18 Hours to Pour the Foundation for San Francisco's Tallest
Skyscraper." Wired.com. November 11, 2015. https://www.wired.com/2015/11/it-
took-18-hours-to-pour-san-franciscos-biggest-ever-concrete-foundation/.
Accessed July 24, 2017.

11 Gretzky, Wayne. Quoted in Brown, Paul B. "You Miss 100% of the Shots You Don't
Take. You Need to Start Shooting At Your Goals." *Forbes*. January 12, 2014.

12 Cowell, Alan. "Sant'Angelo Journal; Corrupt Quake Aid Effort Is Disaster Italian
Style." *New York Times*. April 10, 1993. http://www.nytimes.com/1993/04/10/
world/sant-angelo-journal-corrupt-quake-aid-effort-is-disaster-italian-style.html.
Accessed July 1, 2017.

13 "Kelly, Kerry C. "The Volstead Act." National Archives and Records Administration. Accessed October 26, 2018. https://www.archives.gov/education/lessons/
volstead-act.

14 Pogrebin, Robin. "Neighborhood Report: Harlem; State Pulls Plug on Trade Center."
New York Times. December 17, 1995. http://www.nytimes.com/1995/12/17/
nyregion/neighborhood-report-harlem-state-pulls-plug-on-trade-center.html.
Accessed July 17, 2017.

15 Baglijan, Charles V. "Mayor Claims Credit for Times Sq. Revival." *New York Times*.
Jan 27, 2000. http://www.nytimes.com/2000/01/27/nyregion/mayor-claims-credit-
for-times-sq-revival.html. Accessed June 25, 2017.

16 Cuozzo, Steve. "Captain Cornerstone: If It's a Big Deal in NY Real Estate, Charles
Gargano Had a Hand In It." October 1, 2000. http://nypost.com/2000/10/01/
captain-cornerstone-if-its-a-big-deal-in-ny-real-estate-charles-gargano-has-a-
hand-in-it/. Accessed October 16, 2016.

17 Campbell-Dollaghan, Kelsey. "Celebration, Florida: The Utopian Town That
America Just Couldn't Trust." *Gizmodo*. April 20, 2014. http://gizmodo.com/cele-
bration-florida-the-utopian-town-that-america-jus-1564479405. Accessed April
18, 2017.

18 Cuozzo, "Captain Cornerstone."

19 "Russia: Events of 2016." Human Rights Watch. https://www.hrw.org/world-re-
port/2016/country-chapters/russia. Accessed July 15, 2017.

20 "Russian Federation 2016/2017." Annual Report. Amnesty International. http://
www.amnesty.org/en/countries/europe-and-central-asia/russian-federation/
report-russian-federation/. Accessed July 14, 2017.

21 Yaffa, Joshua. "What the Russian Protests Mean for Putin." *The New Yorker.* March 27, 2017.

22 Bell, Daniel A. "Chinese Democracy Isn't Inevitable." *The Atlantic.* May 29, 2015. https://www.theatlantic.com/international/archive/2015/05/chinese-democracy-isnt-inevitable/394325/. Accessed July 15, 2017.

23 Naim, Moises. "Venezuela: A Dictatorship Masquerading as a Democracy." *The Atlantic.* December 4, 2015. http://www.theatlantic.com/international/archive/2015/12/maduro-venezuela-election-democracy/418860/. Accessed July 15, 2017.

Chapter 4

1 Bush, George W. "The Coalition Against Terrorism." *Vital Speeches of The Day* 68, no. 4: 101. Academic Search Complete, EBSCOhost. Accessed July 19, 2017.

2 Dwyer, Jim and Ford Fessenden. "One Hotel's Fight to the Finish; At the Marriott, a Portal to Safety as the Towers Fell." *New York Times.* September 11, 2002. http://www.nytimes.com/2002/09/11/nyregion/one-hotel-s-fight-finish-marriott-portal-safety-towers-fell.html. Accessed July 17, 2017.

3 Shin, Paul H.B. "Footbridge Near WTC Site Reopens." *New York Daily News,* April 4, 2002.

4 Bush, George W. "Bullhorn Speech." *American Rhetoric: Rhetoric of 9-11.* New York. 2001. http://www.americanrhetoric.com/speeches/gwbush911groundzerobullhorn.htm.

5 Wallach, Ari. "3 Ways to Plan for the (Very)Long Term." TED Video Presentation, 13:42. Filmed October 2016 at TEDxMidAtlantic. https://www.ted.com/talks/ari_wallach_3_ways_to_plan_for_the_very_long_term/transcript. Accessed June 22, 2017.

6 "Core Leadership Values." MasonLeads. 2017. George Mason University. http://masonleads.gmu.edu/about-us/core-leadership-values/

7 Republican National Convention. "Acceptance Speech | President George H. W. Bush | 1988 Republican National Convention." Filmed August 18, 1988. YouTube Video, 31:22–31:49. Posted March 7, 2016. https://www.youtube.com/watch?v=gZCwsEozANM.

8 Goodwin, Doris Kearns. *Team of Rivals: The Political Genius of Abraham Lincoln.* New York/London/Toronto/Sydney/New Delhi: Simon & Schuster. 2005. p. xvii.

9 Janis, I. L. (1982). *Groupthink: Psychological Studies of Policy Decisions and Fiascoes.* Boston: Houghton Mifflin.

10 Abramowitz, Michael. "The Accidental Ambassador." *The American Interest,* 9, no. 3. December 19, 2013. https://www.the-american-interest.com/2013/12/19/the-accidental-ambassador/. Accessed July 14, 2017.

11 Breslow, Jason M. Digital editor. "Colin Powell: U.N. Speech 'Was a Great Intelligence Failure.'" *Frontline.* May 17, 2016. www.pbs.org/wgbh/frontline/article/colin-powell-u-n-speech-was-a-great-intelligence-failure/.

12 Kean, Thomas H. and Lee H. Hamilton. "The 9/11 Commission Report Including Executive Summary: Final Report of the National Commission on Terrorist Attacks Upon the United States." Baton Rouge, LA: Claitors Pub. Division. 2004.

13   Johnson, Richard. "Patricia Lynch Loses in the Sheldon Silver Downfall." *Page Six*. January 23, 2015. http://pagesix.com/2015/01/23/patricia-lynch-loses-in-the-sheldon-silver-downfall/. Accessed April 18, 2017.

14   Williams, Keith. "The Evolution of Hudson Yards: From 'Death Avenue' to NYC's Most Advanced Neighborhood." *Curbed NY*. December 13, 2016. https://ny.curbed.com/2016/12/13/13933084/hudson-yards-new-york-history-manhattan. Accessed July 21, 2017.

15   Campanile, Carl, Lia Eustachewich and Laura Italiano. "Meet the Women Who Allegedly Slept with Sheldon Silver." *New York Post*. April 16, 2016. http://nypost.com/2016/04/16/meet-the-women-who-allegedly-slept-with-sheldon-silver/. Accessed April 18, 2017.

16   Weiser, Benjamin. "Sheldon Silver Appeal Looks to New Definition of Corruption." *New York Times*. March 16, 2017. https://www.nytimes.com/2017/03/16/nyregion/sheldon-silver-appeal.html. Accessed April 18, 2017.

17   Grant, Peter. "Chairman's Board Purge Alters Cablevision Picture." *The Wall Street Journal*. March 2005. https://www.wsj.com/articles/SB110985977098969483. Accessed November 14, 2018.

18   Wilde, Oscar. *Lady Windermere's Fan*. Act 3. 1892.

19   Dupré, Judith. *One World Trade Center: Biography of the Building*. Little, Brown and Company. 2016. p. 110.

20   Daft, Richard. *The Leadership Experience*. 2011. Reprint, Stamford: Cengage Learning. 2015. p. 338.

21   "Northern Manhattan Parks 2030 Master Plan: Public Input and Information Gathering." New York City Department of Parks & Recreation. 2017. https://www.nycgovparks.org/park-facilties/northern-manhattan-parks/master-plan/public-input. Accessed April 18, 2017.

22   Guzi, Glenn. Interview with the ghostwriter, March 29, 2017, OWTC.

23   Gargano, Charles. Quoted in "Lower Manhattan Development Corporation and Port Authority of New York & New Jersey Announce Public Input Process and Timeline Regarding Future of World Trade Center Site and Adjacent Areas." Press Release. May 17, 2002. http://www.panynj.gov/press-room/press-item.cfm?head-Line_id=181. Accessed June 4, 2017.

24   Guzi, Glenn. Interview with the ghostwriter.

25   Fischer, P. et al. The Bystander-Effect: A Meta-Analytic Review on Bystander Intervention in Dangerous and Non-Dangerous Emergencies. *Psychological Bulletin, 137* no. 4. 2011. 517-537.

26   McFadden, Robert D. "Winston Moseley, Who Killed Kitty Genovese, Dies in Prison at 81." *New York Times*. April 4, 2016. https://www.nytimes.com/2016/04/05/nyregion/winston-moseley-81-killer-of-kitty-genovese-dies-in-prison.html. Accessed April 18, 2017.

27   Ellis-Petersen, Hannah. "Boaty McBoatface Wins Poll to Name Polar Research Vessel." *Guardian*. April 17, 2016. https://www.theguardian.com/environment/2016/apr/17/boaty-mcboatface-wins-poll-to-name-polar-research-vessel. Accessed April 18, 2017.

28 Goldberger, Paul. "Miracle Above Manhattan." *National Geographic*. April 2011. https://www.nationalgeographic.com/magazine/2011/04/new-york-highline/. Accessed November 14, 2018.

29 Harrison, Pete. "The Market is Failing the Public, Just Look at The High Line." *Medium*. February 18, 2017. https://medium.com/@petehomeBody/the-market -is-failing-the-public-just-look-at-the-high-line-24c4fe2f6eb3. Accessed May 1, 2017.

30 "The Market is Failing the Public, Just Look at the High Line." HomeBody blog. February 17, 2017. https://joinhomebody.com/blog/2017/2/17/the-market-is-fail-ing-the-public-just-look-at-the-high-line. Accessed April 18, 2017.

31 Gates, Bill. "We Can Eradicate Malaria—Within a Generation." GatesNotes. November 2, 2014. https://www.gatesnotes.com/Health/Eradicating-Malaria-in-a -Generation. Accessed February 7, 2017.

32 "Malaria: Fact Sheet." World Health Organization. Updated December 2016. http:// www.who.int/mediacentre/factsheets/fs094/en/. Accessed February 7, 2017.

33 Sanbrailo, John. "Public-Private Partnerships: A Win-Win Solution." *Pan American Development Foundation*. September 25, 2013. http://www.padf.org/news/ 2014/9/25/huffington-post-public-private-partnerships-a-win-win-solution. Accessed February 7, 2017.

34 "Community Policing Defined." COPS (Community Oriented Policing Services). U.S. Department of Justice. 2012 (rev. 2014). https://cops.usdoj.gov/. Accessed July 19, 2017.

Chapter 5

1 "Seimens Report: Public-Private Partnership Success Stories." NextCity.org. May 7, 2014. https://nextcity.org/daily/entry/siemens-report-public-private-partner-ships. Accessed July 15, 2017.

2 "Five Examples of Public-Private Partnerships (P3) in Action." Onvia.com. October 23, 2013. https://www.onvia.com/company/blog/5-examples-public-private-part-nerships-p3-action. Accessed July 14, 2017.

3 "Examples of Successful Public-Private Partnerships." Sharing Innovative Experiences, 15. Global South-South Development Academy.

4 Guzi, Glenn. Interview with the ghostwriter.

5 Sachs, Jeffrey D. "Why It's Time to Raise the Federal Gas Tax." Politico.com. Jan 19, 2015. http://www.politico.com/magazine/story/2015/01/why-its-time-to-raise-the-federal-tax-on-gasoline-114380. Accessed March 1, 2017.

6 "Increase Corporate Income Tax Rates by 1 Percentage Point." Congressional Budget Office. https://www.cbo.gov/budget-options/2016/52271. Accessed March 1, 2017.

7 Wood, Robert W. "Mark Zuckerberg's $2 Billion Tax Bill Double Last Year, Higher Than Most Billionaires." *Forbes.com*. December 20, 2013. https://www. forbes.com/sites/robertwood/2013/12/20/mark-zuckerbergs-2-billion-tax-bill/#58d5d6745b79. Accessed November 29, 2018.

8 Meyer, David. "Apple Has Paid the $14.3 Billion It Owes the Irish Tax Author-ities—But the Check Hasn't Cleared Yet." September 19, 2018. http://fortune.

com/2018/09/19/apple-ireland-tax-payments-escrow/. Accessed November 25, 2018.

9   Berman, Jillian "8 Outrageous Corporate Tax Breaks." *Huffington Post*. April 30, 2013. http://www.huffingtonpost.com/2013/04/30/corporate-tax-loopholes_n_3179619. html?slideshow=true#gallery/294057/8. Accessed July 25, 2017.

10 Levy, Clifford J. "Port Agency Trims Budget and Its Scope." *New York Times*. July 28, 1995. http://www.nytimes.com/1995/07/28/nyregion/port-agency-trims-budget-and-its-scope.html. Accessed July 13, 2017.

11 Farago, Jason. "One World Trade Center: How New York Tried to Rebuild Its Soul." *The Guardian* (UK). September 8, 2014. https://www.theguardian.com/cities/2014/sep/08/-sp-one-world-trade-center-new-york-rebuild-ground-zero-twin-towers. Accessed March 15, 2017.

12 Shnurer, Eric. "The Secret to Cutting Government Waste: Savings by a Thousand Cuts." *The Atlantic*. July 2, 2013. https://www.theatlantic.com/politics/archive/2013/07/the-secret-to-cutting-government-waste-savings-by-a-thousand-cuts/277458/. Accessed July 19, 2017.

13 Sheehan, Matt. "Beijing's Incredible Subway Expansion in One GIF." *Huffington Post*. http://www.huffingtonpost.com/2014/12/29/beijing-subway-expansion _n_6389002.html. Accessed July 21, 2017.

14 Daley, Jason. "Why You Might Start Seeing Disney And Other Brands in National Parks." Smithsonian.com. May 16, 2016. http://www.smithsonianmag. com/smart-news/why-you-might-start-seeing-disney-other-brands-in-national-parks-180959106/#0aif9Ef7oqbGgtGx.99. Accessed July 14, 2017.

15 McFadden, Robert D. "Derelict Tenements in the Bronx to Get Faked Lived-In Look." *New York Times*. November 7, 1983.

16 Rosenzweig, Roy and Elizabeth Blackmar. *The Park and the People: A History of Central Park*. Ithaca and London: Cornell University Press. 1992. p. 150.

17 Burke, Heather and Bill Arthur. "New Orleans Levees Patched, Army Starts Pumping Water." *Bloomberg*. September 6, 2005. http://archive.is/HXctd. Accessed April 18, 2017.

18 Burke and Arthur, "New Orleans Levees."

19 Schwartz, Frederic. "New Orleans Now—Design and Planning After the Storm." *Natural Metaphor: An Anthology of Essays on Architecture and Nature*. Ed. Josep Lluis Mateo. Zurich: Actar & Eth. 2007. pp. 60–67.

20 Schwartz, "New Orleans Now." p. 63.

21 Jervis, Rick. "For Good or Not, 10 Years of Post-Katrina Rebuilding Changes New Orleans." *USA Today*. August 23, 2015. https://www.usatoday.com/story/news/2015 /08/22/new-orleans-katrina-rebuilding-changes/31417279/. Accessed April 18, 2017.

22 Jervis, "For Good or Not."

23 Long, Colleen. "Nagin Takes Swipe at NYC 9-11 Rebuilding." *The Washington Post*. August 24, 2006. http://www.washingtonpost.com/wp-dyn/content/ article/2006/08/24/AR2006082401668_pf.html. Accessed July 19, 2017.

24 Clark, Patrick and Polly Mosendz. "The Best Way to Influence Congress, According to Staffers." *Bloomberg*. February 13, 2017. https://www.bloomberg.com/news/

articles/2017-02-13/the-best-way-to-influence-congress-according-to-staffers. Accessed April 18, 2017.

25 Wood, Tracy. "A Welcome Addition to OC's Most Park-Poor Neighborhood." *Latino Health Access.* June 6, 2013. http://www.latinohealthaccess.org/2013/06/06/a-welcome-addition-to-ocs-most-park-poor-neighborhood/. Accessed April 18, 2017.

26 Ferguson Publishing. *Career Skills Library: Communication Skills.* 1998. Reprint. New York: Infobase Publishing. 2009. p. 115.

27 Seifman, David. "Gov Blasts WTC 'Traitor'—Claims Greed Sabotaged Talks in Angry Break with Silverstein" *New York Post.* March 16, 2006. http://nypost. com/2006/03/16/gov-blasts-wtc-traitor-claims-greed-sabotaged-talks-in-angry-break-with-silverstein/. Accessed 7-5-17.

28 Seifman.

29 Frangos, Alex. "Silverstein. Port Authority Reach Agreement for World Trade Center." *Wall Street Journal.* April 26, 2006. https://www.wsj.com/articles/ SB114606426334036533. Accessed November 14, 2018.

30 Shapiro, Julie. "Con Edison Made to Share $200 Million it Wanted from 9/11 Fund with Housing, Cultural Groups." DNAinfo. July 29, 2010. https://www.dnainfo. com/20100729/manhattan/conedison-made-share-200-million-911-fund-after-lmdc-vote/. Accessed November 14, 2018.

31 Sykes, Tanisha A. "Retiring at 40 More Than Just a Dream for This Millennial." *USA Today.* June 21, 2017.

32 "Sermon at the Temple Israel of Hollywood." Temple Israel of Hollywood. February 1965.

33 O'Neil, Tyler. "In Name of Civil Rights, Black Dem Senator Disagrees With Martin Luther King, Jr." PJ Media. January 11, 2017. https://pjmedia.com/ trending/2017/01/11/in-name-of-civil-rights-black-dem-senator-disagrees-with-martin-luther-king-jr/.

34 Tarantola, Andrew. "How New York City Built a Massive $3.8 Billion Underground Transit Station in the WTC's Footprints." *Gizmodo.* September 7, 2011. http:// gizmodo.com/5837791/how-new-york-city-built-a-massive-underground-transit-station-in-the-wtcs-footprints. Accessed July 21, 2017.

35 Yee, Vivian. "Fulton Center, a Subway Complex, Reopens in Lower Manhattan." *The New York Times.* November 9, 2014. https://www.nytimes.com/2014/11/10/ nyregion/fulton-center-a-subway-complex-reopens-in-lower-manhattan.html. Accessed September 11, 2018.

36 McCullough, David. *The Path Between the Seas: The Creation of the Panama Canal, 1870-1914.* New York: Simon & Schuster, 1977.

37 Hegeman, Kimberly. "Looking Back on the World's Deadliest Construction Projects." ForConstructionPros.com. August 27, 2015. http://www.forconstructionpros. com/blogs/construction-toolbox/blog/12096401/looking-back-on-the-worlds-deadliest-construction-projects. Accessed June 22, 2017.

38 Smith, Greg B. "Dozens of injuries at World Trade Center construction site went unreported." *New York Daily News.* November 3, 2014. https://www.nydailynews.

com/new-york/exclusive-dozens-injuries-wtc-site-unreported-article-1.1996945. Accessed November 14, 2018.

39 Carter, Shan and Amanda Cox. "One 9/11 Tally: $3.3 Trillion." *New York Times*. September 8, 2011. http://www.nytimes.com/interactive/2011/09/08/us/sept-11-reckoning/cost-graphic.html. Accessed July 11, 2017.

Chapter 6

1   "U.S. Capitol Visitor Center, Over Budget and Late, Set to Open in December." *Fox News*. July 10, 2008. http://www.foxnews.com/story/2008/07/10/us-capitol-visitor-center-over-budget-and-late-set-to-open-in-december.html.

2   Waldman, Benjamin. "The NYC That Never Was: The George Washington Bridge Was Supposed To Be a Beaux Arts Masterpiece." *Untapped Cities*. October 20, 2013. http://untappedcities.com/2013/10/30/nyc-that-never-was-george-washington-bridge-beaux-arts/. Accessed June 20, 2017.

3   McCarthy, Niall. "Major International Construction Projects That Went Billions Over-Budget" [Infographic]. *Forbes*. December 10, 2014. https://www.forbes.com/sites/niallmccarthy/2014/12/10/major-international-construction-projects-that-went-billions-over-budget-infographic/#406dae45376a. Accessed June 24, 2017.

4   "The Apollo 13 Accident." NSSDCA/NASA. https://nssdc.gsfc.nasa.gov/planetary/lunar/ap13acc.html. Accessed June 22, 2017.

5   Chow, Denise. "What If Apollo 13 Failed to Return Home? New Video Tells All." SPACE.com. April 13, 2010. https://www.space.com/8203-apollo-13-failed-return-home-video-tells.html. Accessed June 22, 2017.

6   Crass, Stephen. "Apollo 13, We Have a Solution." *IEEE Spectrum*. April 1, 2005. http://spectrum.ieee.org/aerospace/space-flight/apollo-13-we-have-a-solution. Accessed June 19, 2017.

7   "New York Lays Cornerstone for Freedom Tower." *CNN*. July 6, 2004. http://www.cnn.com/2004/US/Northeast/07/04/wtc.cornerstone/. Accessed July 14, 2017.

8   Sagalyn, Lynn. *Power at Ground Zero: Politics, Money, and the Remaking of Lower Manhattan*. New York: Oxford University Press, 2018. p. 276.

9   *Daily News*. "Get New York Back on Track." May 6, 2005.

10 Healy, Patrick D. and William K. Rashbaum. "Security Concerns Force a Review of Plans for Ground Zero." *New York Times*. May 1, 2005. http://www.nytimes.com/2005/05/01/nyregion/security-concerns-force-a-review-of-plans-for-ground-zero.html. Accessed July 20, 2017.

11 Candela. "Address to the Bishop, Mayor, and Charles Gargano." The Church of St. Anthony's Martyr, Bishop's residence. August 12, 2007.

12 "Here's How Much Lower Manhattan Has Changed Since the 9/11 Attacks." *Fortune*. September 9, 2016. http://fortune.com/9-11-ground-zero-lower-manhattan. Accessed July 14, 2017.